THE FDA FOLLIES

THE
FDA FOLLIES

HERBERT BURKHOLZ

BasicBooks
A Division of HarperCollinsPublishers

Material contained in chapters 1 and 2 appeared, in different form, in "A Shot in the Arm for the FDA," *New York Times Magazine,* June 30, 1991. Material in chapter 6 appeared in "To B12 or Not To B12," *Lears,* April, 1993; different material from chapter 6 appeared in "Killer Grapes," *New Republic,* November 30, 1992. Material from chapter 9 appeared in "Bad Blood?," *Town & Country,* March, 1992.

Copyright © 1994 by Herbert Burkholz.
Published by BasicBooks, A Division of HarperCollins Publishers, Inc.

Library of Congress Cataloging-in-Publication Data
Burkholz, Herbert, 1934–
    The FDA follies / Herbert Burkholz.
        p.   cm.
    Includes bibliographical references and index.
    ISBN 0-465-02369-X
    1. United States. Food and Drug Administration. 2. Food adulteration and inspection—United States. 3. Drugs—Inspection—Government Policy—United States. 4. New products—Inspection—Government policy—United States. I. Title.
HD9000.9.U5B87   1994
363.19'2'0973—dc20   93-32913
                                                                    CIP

94 95 96 97 ❖/HC 9 8 7 6 5 4 3 2 1

For Ron and Debbie Sauder

There's no trick to being a humorist when you have the whole government working for you.

—Will Rogers

CONTENTS

ACKNOWLEDGMENTS

My INTEREST IN THE FOOD AND DRUG ADMINISTRATION WAS FIRST sparked by my friend and colleague, Robert Teitelman, during the time when he was writing about pharmaceutical companies as a magazine journalist. The tales that he told—horror stories, some of them—prompted me to take a long, hard look at this agency so vital to the national welfare. I was not happy with what I saw, and the next step was an assignment from the *New York Times Magazine*, and my editor, Katherine Bouton, to write about the FDA. The result was a cover story for the magazine, but another, unexpected, result was the realization that there was a great deal more to be told, far more than could be crammed into the narrow confines of a magazine article. Thus, this book.

There are several people to thank, those without whom the making of this book would have been far more difficult. My thanks are due to Dr. Sidney Wolfe for sharing his insights with me, and to David Nelson, the former chief investigator for the Subcommittee on Oversight and Investigations of the House Committee for Energy and Commerce. Also associated with that committee were Reid Stuntz, the chief of staff, Claudia Beville, Steven Simms, and Tom Dornay. Mitchell Zeller, who at that time was counsel for the

House Committee on Human Resources, provided helpful advice and information, as did attorneys Bruce Finzen of Minneapolis and David Holzworth of Washington, D.C.

Dr. David Kessler was at all times open and forthcoming with me.

My thanks are due to the librarians of the FDA Medical Library, the National Library of Medicine in Bethesda, and the Montgomery County, Maryland, public library system, which, pound for pound, ranks with the best.

And there are those who cannot be thanked by name: those men and women of the FDA who spoke to me under the cloak of anonymity.

Finally, my thanks to Susan Burkholz, not the customary tip of the hat from the author to his wife, but a sincere acknowledgment of the incomparable job of research that she did for this book, just as she has done for my books in the past.

1

A Flock of Rejected Suitors

IF YOU HAVE ANY SYMPATHY FOR THOSE WHO WORK IN GOVERNMENT, THEN spare a tear or two for the public servants who labored at the Food and Drug Administration (FDA) during the 1980s. The decade was a terrible time for them. Like a flock of rejected suitors, they wanted desperately to be loved. They yearned for the respect of the nation, and they knew that they did not have it. Charged with the mission of assuring the purity, safety, and efficacy of our foods, our medicines, our blood supplies, and even our cosmetics, they envisioned themselves as the guardians of the national health, and they desperately wished that the rest of the country would see them that way too.

But for most Americans during the 1980s, the FDA was perceived as bumbling and inefficient, a target under constant attack from congressional committees, newspaper editorials, public interest groups, and the executives of the very industries it was trying to regulate. It was seen as the epitome of a foot-dragging bureaucracy, the agency that took thirty years to write the regulations for over-the-counter drugs and another thirty years to write the regulations for color additives in foods; the agency that enforced its regulations in so casual and haphazard a manner that industry compliance was as much a matter of chance as it was of law. It was an agency singed by scandal, hobbled by outmoded regulations and procedures, and enfeebled by a leadership that lacked both verve and imagination. It was an agency that seemed to shoot itself in the foot with a relentless regularity, and always just in time for the six o'clock news. It was an agency that so exasperated its critics that to the staff of one congressional committee it was a standing joke that, "When dealing with the FDA, one should never suspect dishonesty as long as ineptitude is still a possibility."

Because of this, covering the FDA during the 1980s was a journalist's dream and a nightmare combined. It was a dream because it was the ongoing sort of story that seems never to die. Something was always happening at Parklawn, the FDA headquarters in Rockville, Maryland. One day it was a new accounting of administrative sloth, the next a spicy stew of corruption in the ranks, perhaps a salvo of congressional criticism from the Hill, or a long-range bombardment from AIDS (acquired immunodeficiency syndrome) activists all over the world. . . . It never stopped. But it was a nightmare because of the agency's defensiveness about any controversial subject, a circle-the-wagons and take-no-prisoners attitude that admitted to no imperfections in the FDA facade.

Aside from the routine handouts, getting information from an FDA press officer was an exercise in futility. Calls were rarely returned, off-the-record comments were discouraged, and known critics of the agency were treated as pariahs. Most of this was prompted by nothing darker than a deep sense of loyalty and a conviction that the agency was being poorly used, but the attitude came across as an arrogant, damn-your-eyes indifference to public opinion. Working under those conditions, it is not surprising that the FDA was presented to the public by the printed and the electronic media as the ultimate bureaucracy. And not without reason if a bureaucracy is to be defined as a system of administration marked by officialism, red tape, and self-serving job protection.

But the FDA-ers had a valid complaint about being poorly used, for the agency was in no way singular in its bureaucratic bent. Government service has always attracted those for whom the demands of the private sector are too stringent, those for whom responsibility is too heavy a burden, and those who are devoid both of imagination and ambition. And government service will always attract that breed of man and woman to whom employment in the public interest is the highest form of good, and to whom salary will always be second to satisfaction. Bureaucracy may appear to be a monolith, but it never is. It is always an uneasy coalition composed in varying degrees of the daring and the meek, and as such it exists in all parts of government. The true bureaucrats in the Commerce Department, or Treasury, or State, are certainly the equals of their FDA counterparts in terms of timeserving, paper shuffling, and general obfuscation. The difference between the FDA and these others lies in the public perception of their functions, for the FDA is responsible for nothing less than the health of the nation, and that national health is too precious to be left in the hands of bureaucrats.

One certainly might question if national health is so much more precious than our treasury, our commerce, our national defense, all of which are served by men and women of the same bureaucratic bent. Perhaps not, but somehow the American public has perceived it that way. Treasury, and commerce, and national defense are abstract concepts to most Americans, vitally important . . . but not personal. The purity of the foods we eat and the safety of the medicines that we take are concerns that are woven into the very fabric of our everyday lives. There is nothing abstract about those concerns, and to the public eye the journalistic parade of FDA horror stories was a source first of amusement, then of disbelief, and finally of fear. No one, it seemed, had a good word to say about the FDA.

To the rejected suitors at the FDA, much of this public perception seemed grossly unfair, and based on the sort of high expectations that can never be satisfied in an imperfect world, but in their heart of hearts they also knew that there was something very wrong with the FDA of the eighties. Something was so obviously wrong that as the decade drew to a close in March 1990, the secretary of the Department of Health and Human Services (HHS), Dr. Louis W. Sullivan, announced the appointment of an advisory committee to study the agency's shortcomings. The committee's report, published just over a year later, was damning both in general and in detail.[1] The panel of fifteen experts, headed by Dr. Charles C. Edwards, himself a former FDA commissioner, found the FDA's laboratories and equipment to be in abysmal condition.

It found food factories that had been inspected only once in eight years.

It found drug factories that had not been properly examined because of a shortage of inspectors.

It found the agency overwhelmed and incapable of coping with the vastly increased duties caused by the AIDS epidemic.

It found an agency operating on a threadbare budget, close to impotence, and badly in need of expanded powers to issue subpoenas to seize products, and to impose monetary penalties on companies that violated regulations.

It found an agency that no longer had the scientific ability to evaluate new drugs, or to keep up with the revolutionary advances occurring in the biological and medical sciences.

It found the potential for a national disaster in the making.

From aspirin to AIDS drugs, from soup to fish, from heating pads to heart valves, 25 cents out of every dollar spent by the American consumer goes for products regulated by the FDA. For health-conscious Americans, already concerned to the point of obsession with the foods they were eating and the medicines they were taking into their bodies, the Edwards report was a sobering reminder that no one needed overseeing more than those who oversaw the national health.

There was one further weakness in the FDA that the Edwards report did not address, and that was the institutional malaise that had colored the agency's actions during the 1980s. When the FDA relaxed the enforcement of regulations at the outset of the Reagan administration, it began to lose control of the marketplace. As early as 1982, Vice-President George Bush was able to say to a group of representatives from the drug industry that, "I think we've started to see this philosophical shift, the end or the beginning of the end of this adversarial relationship. Government shouldn't be an adversary. It ought to be a partner."[2]

On the surface, it was a point that was difficult to contest. Under ideal circumstances, the FDA and the industries that it regulates should have the same goals: the production of safe and effective drugs, and the distribution of clean and wholesome foods. But to join industry and government in partnership is to invite the lion to lie down with the lamb. Sometimes it works, but all too often the result is lamb stew for lunch, and in the 1980s there was lamb on the menu almost every day.

To those businesspeople who have to deal with regulatory agencies, the concept of industry as the lion and government as the lamb seems laughable. They see the situation as quite the opposite, with government as the intrusive predator, harassing and intimidating both the honest and the deviant with fine impartiality. But unlike other regulatory agencies, the FDA enjoys a power over the industries it regulates that may be clearly defined by statute, but is unenforceable by statute alone. The function of the agency, first and always, is to protect the public health by *regulation*, which often means to restrict, to forbid, and, if necessary, to punish. It is a job for a cop, no matter how much gloss is put on the word, but regulating the massive industries that supply the nation's food and drugs requires something more than a policeman's vigilance, and a great deal more than a policeman's justice.

It requires the implicit cooperation of the industries under regulation, and the history of the FDA may be seen as an ongoing attempt on the part of the agency to secure that cooperation, and an ongoing reluctance on the part of industry to grant it. Industry pays lip service to the need for cooperation, but it takes meaningful action only when a public crisis, a public tragedy, or an outraged public opinion makes a lack of compliance unthinkable. The contest between private profit and public health in America was joined on the day when the first colonial farmer sold his first quart of adulterated milk, when the first pint of snake oil was peddled over the back of a buckboard, and it has been going on ever since. Only the forms of the contest change in varying shades of sophistication, but the substance is always the same. How little regulation can be used yet still be effective? How much regulation is too much of an intrusion into the affairs of a private enterprise?

It is difficult for us, as we approach the end of the twentieth century, to imagine the conditions that prevailed in the food and drug industries at the end of the nineteenth. As America changed from an agricultural to an industrial economy, the feeding of an ever-growing urban population became an imperative that required the transportation of food over longer and longer distances. But sanitation was primitive, ice was the principal means of refrigeration, and the use of chemical preservatives and toxic colors was virtually uncontrolled.

In that same era, thousands of home-brewed patent medicines with names like "Kick-a-poo Indian Sagwa" and "Warner's Safe Cure for Diabetes" reflected both the limited medical knowledge of the times and the ingrained doctrine of caveat emptor. Opium, morphine, heroin, and cocaine were laced into medicines and sold unlabeled, and without restriction. Ingredients were unlisted, warnings against misuse were unheard of, and innocuous preparations were labeled as the cure for every disease from dandruff to cancer. If the public learned anything at all, it learned only from bitter experience.

The first general law against food adulteration in the United States was enacted by Massachusetts in 1784, but Wallace F. Janssen, the noted historian of the FDA, records that as far back as

1630 the Massachusetts Bay Colony sentenced one Nicholas Knopf to pay a fine or be whipped for selling "a water of no worth nor value" as a cure for scurvy.[3] Gradually, other states passed a variety of food and drug statutes, but as the country expanded it became clear that a national law was needed. Many states had no laws at all, others lacked the power to enforce their feeble statutes, and products that met the requirements of one state could well be judged illegal in the next. During the first half of the nineteeth century the various states renewed such laws sporadically, but the national government carefully avoided either posing or answering the question of federal responsibility and jurisdiction.

It was inevitable that such catch-as-catch-can regulation would result in a fading of faith in medicines in general, and that so highly respected a physician as Dr. Oliver Wendell Holmes, professor of anatomy at Harvard Medical School, could say with conviction in 1860: "Throw out opium, . . . throw out a few specifics which our [physicians'] art did not discover, and it is hardly needed to apply; throw out wine, which is a food, and the vapors which produce the miracle of anaesthesia, and I firmly believe that if the whole materia medica, as now used, could be sunk to the bottom of the sea, it would be all the better for mankind—and all the worse for the fishes."[4]

It was not until the Mexican War (1846–1848), when a crisis developed over medications for the troops, that Congress enacted a law that banned adulterated drugs from being offered for import. It took three more decades before the Congress even began to consider a general food and drug measure, and another thirty years before the landmark law of 1906 was enacted. In the years prior to enactment, the first advocates of federal intervention were the state officials who knew quite well the problems and the weaknesses of existing controls. But it was the leadership of one remarkable man, Harvey Washington Wiley, that finally made food and drug protection a function of the federal government.

In 1883, Dr. Wiley left Purdue University to become chief of the Bureau of Chemistry of the U.S. Department of Agriculture, and one of his first official acts was to expand studies of food adulteration, thus planting the seed of what would one day become the Food and Drug Administration. Wiley's chemists had no difficulty in obtaining evidence of widespread adulteration for their investigations and reports, and Wiley took their findings to the public. He became a popular speaker at women's clubs and business organizations, and a group of crusading writers joined in the campaign. National magazines such as *Collier's Weekly, Ladies Home Journal,* and *Good Housekeeping* aroused public opinion with their cartoons, articles, and editorials.

Strenuous opposition to Wiley's campaign for a federal food and drug law came, not unexpectedly, from the makers of patent medicines and the many food manufacturers, and that campaign might have ended in failure, at least for the moment, had the nation not been aroused by the publication of a single book. In his avowedly socialist novel, *The Jungle,* Upton Sinclair so graphically described the filthy conditions under which the nation's meat supply was prepared that a wave of public indignation rolled over the country, meat sales dropped by half, and an angry President Theodore Roosevelt helped to push both a meat-inspection bill and the food and drug law through Congress.

Administration of the new law was assigned to the Bureau of Chemistry, and the young men and women recruited by Wiley and his successors developed an efficient organization, winning scores of judicial interpretations that led both to a strengthening of the law and an examination of its weaknesses. The Bureau of Chemistry enforced the 1906 law until 1927 when the Food, Drug, and Insecticide Administration was formed, to be renamed in 1931 as the Food and Drug Administration. In 1940, the FDA was transferred from the U.S. Department of Agriculture to the Federal Security Agency which, in 1953, became the Department of Health, Education, and Welfare, and then the Department of Health and Human Services.

The law, as enacted in 1906, was strong for its time, but was filled with loopholes. False therapeutic claims for patent medicines were rarely prosecuted because the burden of proof fell on the government to show that the manufacturer had intended to defraud its victims. Thus, the defendant had only to show that he personally believed in his product in order to escape prosecution. Food adulteration continued to flourish because judges could find no specific authority in the law for the standards of purity and content that the FDA had established, and the economic hardships of the depression years served to underscore the many shortcomings of the original act.

In 1933, a few days after the inauguration of President Franklin D. Roosevelt, the FDA chief, Walter Campbell, approached Rexford Tugwell, a member of the president's "brain trust," and the new assistant secretary of agriculture, with a plan for the revision of the Food and Drug Act. Nothing seemed impossible in those heady early days of the first Roosevelt administration, and the "Tugwell Bill" was quickly slapped together. It was a legislative disaster, vehemently opposed by industry and advertising interests, and fated from the start for certain defeat. It was left to Senator Royal S. Copeland of New York, himself a medical man, to pick up the pieces and fashion a bill that had a chance of passing, while still providing essential consumer protection.

But nothing was accomplished until, as had happened in the past and would happen again in the future, disaster forced the event. In July 1937, the chief chemist of the Massengill Company, searching for a palatable solvent for the new wonder drug sulfanilamide, used an ethylene glycol/water mixture flavored with raspberry extract. In September, 240 gallons of the product, "Elixir Sulphanilamide," were shipped to pharmacies, and shortly thereafter people started dying. Distribution was stopped, and a frantic search for bottles of the elixir began. By the time that the last bottle had been recovered, 107 people had died, poisoned by the medicine. Massengill had not tested its product on animals prior to marketing, but rather had relied on the knowledge that ethylene glycol, as such, was safe at the dosage used, and that sulfanilamide was known to be a safe and potent drug. What

Massengill had not known was that these two purportedly safe ingredients were highly lethal in combination.

Within months of the "Elixir" disaster, Congress finally passed the Federal Food, Drug, and Cosmetic Act, replacing the original legislation. It became law on June 30, 1938. Although many compromises had to be made to secure passage, the new law was a major improvement:

Cosmetics and therapeutic devices were regulated for the first time.

Proof of fraud was no longer required to stop false claims for drugs.

Drug manufacturers were required to provide scientific proof that new products could be safely used before putting them on the market.

The addition of poisonous substances to foods was prohibited except where unavoidable or required in production. Safe tolerances were authorized for residues of such substances, as in the case of pesticides.

Specific authority was provided for factory inspections.

Food standards were required to be set up when needed "to promote honesty and fair dealing in the interest of consumers."

Federal court injunctions against violations were added to the previous legal remedies of product seizures and criminal prosecutions.

The new law, and World War II, expanded the FDA's workload, particularly in the testing of the new "wonder drugs," but consumers continued to be human guinea pigs for a host of chemical compounds of unknown safety. The agency could and did stop the use of many known poisons, but the research effort needed to prove that all food chemicals were safe was clearly beyond its resources. In 1949, Commissioner Paul B. Dunbar took the problem to Congress, and the result was a series of hearings on the dangers of chemicals in food, chaired by Representative James T. Delaney of New York. From these hearings came three amendments that fundamentally changed the character of our food and drug laws: the Pesticide Amendment of 1954, the Food Additives Amendment of 1958, and the Color Additive Amendments of 1960.

With these laws on the books, it could be said for the first time that no substance could legally be introduced into the U.S. food supply unless there had been a prior determination that it was safe. By placing the onus of research on the manufacturers, a problem of unmanageable size had been made manageable, and by preventing violations through premarketing clearance procedures, the consumer had been given immeasurably better protection than before, when violations could be proven only after injuries had been reported.

The trend toward a preventive law continued, and the thalidomide tragedy in Europe, in which thousands of deformed infants were born, helped to focus public attention on pending legislation designed to strengthen the federal Food, Drug, and Cosmetic Act. The Drug Amendments of 1962, passed unanimously by Congress, tightened control over prescription drugs. It was recognized that no drug is truly safe unless it is also effective, and effectiveness was required to be established prior to marketing. Drug firms were required to send adverse reaction reports to the FDA, and drug advertising in medical journals was required to provide complete information to the doctor—the risks as well as the benefits. In the years after the passage of the 1962 amendments, literally thousands of prescription drugs were taken off the market because they lacked evidence of both safety and efficacy, and thousands more were forced to change their labeling claims.

Although the Food, Drug, and Cosmetic Act of 1938 has served the nation well, it requires constant revision in order to keep it abreast of current scientific, economic, and social realities. Hardly a year passes without some change. In 1976, an amendment strengthened the FDA's authority to regulate medical devices, in the same year another amendment prevented the agency from establishing standards for vitamin and mineral food supplements, and in 1977 the agency was constrained from acting to stop the use of saccharin in foods and beverages.

In 1980, serious illness in babies due to a mineral deficiency caused Congress to pass an Infant Formula Act requiring strict controls to ensure the nutritional content and safety of commercial baby foods. Two years later, in 1982, seven deaths from cyanide poisoning traced to Tylenol pain-relief capsules caused the FDA to develop and issue regulations requiring tamper-resistant packaging, and Congress passed a law making it a crime to tamper with packaged consumer products.

Rare diseases often lack effective treatment because pharmaceutical manufacturers are reluctant to invest in costly research for a product without a marketing future, but a new approach to this problem was found with the passage of the Orphan Drug Act of 1983. Under this law, FDA was given the power to designate products that are eligible for tax credits covering the costs of clinical trials, and exclusive market rights were given to the maker for seven years if the drug was not patentable.

In 1984, the Drug Price Competition and Patent Term Restoration Act was passed, permitting FDA to approve generic versions of previously approved drugs without requiring the sponsors to duplicate the costly human tests that were required for the originals. It also allowed the usual seventeen-year term of a patent on a medicine to be extended by up to five years in order to compensate the manufacturer for the time required in getting FDA approval.

With the passage of the original law and its continuing amendments, the federal government confirmed the need of people to protect themselves from the commercial adventures of others, and the history of the FDA has been the history of an agency devoted to that need. By statute, and by its actions over the years, the FDA eventually came to represent the best that there was in public service: high, but attainable, scientific goals that were sought by dedicated professionals, and by the end of the 1970s the agency enjoyed a worldwide respect as the finest regulatory body of its kind. It was the model on which all others were based, the source of wisdom for sister agencies in Great Britain, Germany, Canada, and Japan. It was a proud place, peopled by proud men and women who saw their lives as part of a distinct and laudatory purpose.

What happened then, in the 1980s, to transform that proud place into a symbol for bumbling bureaucracy, a byword for inefficient foot-dragging, and a fertile field for greed and corruption? Did some unknown force throw a master switch that could subvert ideals overnight, dilute initiative, sap enthusiasm, and turn the best of its kind into a sorry caricature of itself? Clearly not. The seeds that came to fruition in the 1980s had been sown years before, the benign intent of the agency's power had been bending toward autocracy for decades, and the creeping lassitude that came to inhibit scientific progress was nothing more than the bureaucratic ennui that can afflict even the most devoted of public servants. But something did happen at the beginning of the decade, something that crystalized these flaws along a fault line, and that produced the FDA Follies of the 1980s. The seeds were there, but they started to sprout when Ronald Reagan and George Bush invited the lion to lie down with the lamb.

Over the years, including the 1980s, the two most vocal and effective critics of the lion and the lamb have been a congressman and a consumer advocate. The congressman, Representative John Dingell (D-Mich.), has given the FDA little peace in his position as chairman of the House Committee on Energy and Commerce and its Subcommittee on Oversight and Investigations. The savage sessions of his subcommittee hearings have been likened to the arenas of ancient Rome, where errant FDA officials are grilled, roasted, and thrown to the wild beasts; and his seemingly endless stream of critical letters to FDA commissioners gave birth to the term "dingellgram" at agency headquarters in Rockville. On any controversial subject involving the FDA, John Dingell can usually be found at the center, and any such subject dealt with in these pages is bound to bear his imprint.

The consumer advocate, Dr. Sidney Wolfe, has been called a stinging gadfly on health issues, an armored crusader, a Jeremiah crying out in the wilderness, and a Joan of Arc with a singularly clear view of what is right about American medicine, and what is very wrong. For more than two decades, Wolfe has led the nonprofit Public Citizen Health Research Group in hundreds of public health crusades, including successful drives to ban red dye no. 2, the diabetes drug phenformin, and other drugs for high blood pressure and arthritis. Wolfe cofounded Public Citizen Health with Ralph Nader in 1971, and since then his spare office on Washington's Dupont Circle has issued more than 1,200 reports criticizing various drugs, devices, foods, and medical practices. He led the battle with the FDA over the the Bjork-Shiley heart valve, silicone breast implants, and toxic shock syndrome related to the use of tampons. During the 1980s he was directly responsible for goading, pressuring, and eventually bringing legal action against the FDA to force the use of warning labels against Reye's syndrome on aspirin bottles. In the Reye's case, as he has many times, Wolfe took on an avaricious industry, a business-oriented administration, and a supine FDA. Eventually, he won, but not before lives were unnecessarily lost.

Many of those lives were lost because of the Reagan administration's refusal to antagonize the aspirin industry by insisting on warning labels on aspirin bottles, labels that would advise the public of the dangers of Reye's syndrome. In 1982, when the connection between aspirin and Reye's was clear and dangerous, Wolfe said in a position paper that, "If the Reagan administration continues to stutter on this issue, the number of children who will die or get nerve damage will make the Tylenol disaster look microscopic in comparison. Instead of a deranged person lacing Tylenol with cyanide, the Reagan administration . . . , backed by the aspirin industry and the misguided American

Academy of Pediatric Executive Board . . . , is lacing mothers and fathers with misinformation downplaying the link between aspirin and Reye's syndrome."[5]

The statement was vintage Wolfe. It was blunt, irreverent, uncompromising, and accurate. The condition that we now know as Reye's syndrome was observed as far back as 1929, but it was first identified and characterized as a distinct entity during the 1950s at the Royal Alexandra Hospital for Children in New South Wales, Australia. Each year between 1951 and 1962, the hospital noted admitting a small number of children in such a critical state that most of them could not be saved. When admitted, almost all were in a coma or stupor, although their illnesses had started only a few days earlier with only common childhood respiratory symptoms. Seventeen of the children died within an average of twenty-seven hours after admission, and, at autopsy, all were found to have brain swelling, a bright yellow liver, and kidney damage.

Dr. Douglas Reye, the hospital's director of pathology, believed that this set of symptoms represented a distinct disease, although of unknown cause. Then, in 1963, Dr. George Johnson reported a similar set of sixteen fatal cases during an outbreak of influenza B in North Carolina, and after that the condition was known as the Reye-Johnson syndrome, although usually shortened to Reye's.[6] Investigators looking for some common factor among the children who developed the syndrome found it in aspirin, which had been taken during flu or chicken pox, and in 1980, the results of studies conducted in Ohio, Michigan, and Arizona demonstrated an association between Reye's and the use of aspirin during a preceding respiratory tract or chicken pox infection. The association was confirmed in 1981 at a conference called by the Centers for Disease Control (CDC) in Atlanta, where evidence was presented to show a strong relationship between the use of aspirin and the development of Reye's syndrome. This evidence was sufficient for Wolfe to petition the FDA for warning labels on all aspirin-containing products.

"The failure of FDA to do so," he said at the time, "aside from violating the drug laws which require such labeling in the face of the evidence of the dangers of aspirin, will guarantee the continued deaths and injuries to children."[7] Sadly, the children continued to die. When the Department of Health and Human Services (HHS), the parent body to the FDA, decided that warning labels on aspirin bottles were necessary to protect the health of the nation's children, that decision was at once reversed under heavy pressure from the aspirin industry. Always sensitive to the needs of American industry, the Reagan White House would do nothing more than mildly advise parents through public service announcements and information brochures to be cautious in the use of aspirin in treating children for chicken pox or flu.

Yet even this halfhearted effort came under fire from the aspirin industry, which threatened to sue anyone who disseminated the educational material. Then, aware that it was painting itself into a public relations corner, the industry counterattacked with the formation of a front organization called the American Reye's Syndrome Association, a group partially funded by the makers of Bayer Aspirin for children. The ostensible purpose of the organization was to promote the diagnosis and treatment of Reye's syndrome, but its actual purpose was to provide the industry with a cosmetic shell while it continued to fight against warning labels. The issue of warning labels was the one nonnegotiable item that the industry had to defend. Any other point could be sacrificed, like throwing children from sleds to the wolves, in order to prevent labeling.

Thus, the newly formed association set about distributing pamphlets that bore such engaging titles as "Because You Love Children, Learn About Reye's Syndrome." In these carefully crafted pieces of propaganda, parents were told how to check for symptoms of Reye's, and were supplied with a list of do's and don'ts, among which was the hedging suggestion that if medication (read aspirin) was required to reduce fever, it "should only be used as needed to keep the fever in control." Even while indirectly suggesting that aspirin should not be used in these circumstances, nowhere in the association's literature was there even a hint of the direct connection between Reye's and aspirin, despite the fact that in the twenty-month period following the original CDC recommendation, an additional 328 cases, including 99 deaths, had been reported, and many more were thought to have gone unreported.[8]

By October 1983, those figures had risen to 361, including 113 deaths, and no scientist knowledgeable on the subject still doubted the association between aspirin and Reye's. From the surgeon general's office to the CDC, officials were advising physicians and parents of the risks involved, and even an FDA working group was able to state that the association of "salicylates [aspirin] and Reye's syndrome . . . was sufficently strong to warrant warning health professionals and parents."[9]

It was at this point, in 1983, that the secretary of HHS, Margaret Heckler, in a rare display of independence, ordered the FDA to arrange for the printing of half a million consumer pamphlets to be distributed in 4,200 supermarkets across the country, warning parents about the association between aspirin and Reye's syndrome. The FDA was also instructed to schedule for distribution to 5,000 radio stations a thirty-second-spot warning against using aspirin in the treament of chicken pox and flu. While falling short of the warning label, which was the goal of consumer advocates such as Wolfe, the project was intelligently conceived, constructive, workable, and impossible to implement during the Reagan years. In early October, after a meeting between Heckler, White House officials, and representatives of the aspirin industry, HHS reversed itself by 180 degrees and ordered the project dropped. The pamphlets were never distributed, the spot was never aired, and the children kept on dying.

"By banning the distribution of these educational materials," said Wolfe at the time, ". . . the Reagan administration is willing to condemn American children to death or brain damage from Reye's syndrome in order to please the aspirin industry."[10] At this point it would have been a rare public health official who would have denied, in private, the connection between Reye's and aspirin, and Wolfe touched on the irony of the situation on November 5, 1984, in one of his many letters to FDA Commissioner Frank Young. "As parents," he noted, "neither you, nor I, nor our wives would give aspirin to our children if they had chicken pox or flu. As FDA commissioner, I hope you will enable all other American parents to make the same safe choice by denouncing the misleading message of the aspirin industry–funded group."[11]

Wolfe kept up the heat over the years, calling in vain upon the FDA to insist on warning labels, and denouncing every countermove made by the aspirin industry. When a so-called Aspirin Foundation promised to put warning signs in stores voluntarily, Public Citizen Health ran a survey and found that, of the pharmacies visited, less than one third had posted the signs. When the Committee on the Care of Children (CCC), another aspirin industry front group, placed misleading and dangerous "public service" announcements about Reye's with television and radio stations nationwide, Public Citizen Health wrote to over 1,000 of those stations, urging them not to use the announcement. Whenever he could, Wolfe campaigned for warning labels, and in 1984 Public Citizen Health went into the U.S. District Court (D.C.) to litigate the issue of unreasonable delay in requiring warning labels for aspirin products.[12]

Finally, on March 15, 1985, Wolfe appeared before the House Health Subcommittee chaired by Representative Henry Waxman (D–Calif.) to urge the passage of H.R. 1381, a bill to require warning labels, advertisements, and store signs concerning aspirin and Reye's syndrome. His testimony was blunt, reiterating the facts and the accusations he had been making for years, and on October 29, 1985 H.R. 1381 was introduced, and was subsequently enacted into law. In 1986, the FDA at last issued its regulations making warning labels on aspirin products mandatory.

The Reye's syndrome episode was a tragedy for American public health not because of its size, not because of its impact on the nation, but because of the cold-blooded business-as-usual attitude of the Reagan White House that brought it to pass. By the time that Sidney Wolfe testified before the Waxman Committee, 610 children had been stricken, 175 of them fatally, since the CDC recommendation on October 14, 1981, that the use of aspirin "should be avoided" for the treatment of chicken pox or flu. The Reye's episode was a typically American tragedy in that a society driven by commerce, and supported by a business-oriented administration, found no difficulty in opting for profits over the lives of a few hundred kids. As for the FDA, like the rest of HHS, it shamelessly rolled over and played dead for the aspirin industry, thus epitomizing the actions of the agency for the decade. In its best defense, one can only assume that these were men and women who were acting in fear for their jobs, but one has to wonder how Dr. Harvey Washington Wiley would have acted in the same situation.

In her remarkable book, *The March of Folly*, Barbara W. Tuchman points out that, "A phenomenon noticeable throughout history regardless of place or period is the pursuit by governments of policies contrary to their own interests."[13] This she defines as institutional folly, asking why holders of high office so often act contrary to the way reason points and enlightened self-interest suggests. Tuchman chooses four historical examples of the phenomenon, ranging from the episode of the Trojan horse to the American involvement in Vietnam, and in each of them she shows the almost unlimited capacity of government, be it composed of princes or of popes, to blind itself to reason and to embrace a needless disaster.

In a similar sense, the workings of the FDA during the 1980s seem to defy understanding. It would have been both logical and prudent for the FDA to concern itself with the spiraling cost of pharmaceuticals in the 1980s. Traditionally, drug companies had tried to excuse high prices on the grounds of the high cost of research, but by the end of the decade the industry was spending more on the promotion and marketing of its products than it was on research. Since promotion was designed to create a market for a product, and the market that was created was not necessarily the market that would benefit from the drug, the end result often was the inappropriate prescription by physicians. This was a matter of clear concern to the FDA, but the problem of escalating drug prices was ignored and left for another generation to solve.

It was clearly in the interests of the FDA to maintain its authority in the enforcement of its mandate, but all too often the agency ceded that authority, or declined to exercise it. It was clearly in the interests of the agency to punish incompetence within its ranks, and to reward diligence, but all too often the opposite occurred.

It would have been eminently logical for the FDA to take a leading role in the fight against AIDS, but at first the agency actually questioned the existence of the epidemic, and it later used procedures totally out of tune with the times to delay the approval of drugs sorely needed. It would have been equally logical for the FDA to regulate strictly the safety of the nation's blood supply, but in the beginning the agency dithered and people died.

It was clearly in the best self-interests of the agency to root out internal corruption. But it took a pair of whistle-blowers and a congressional investigation to force the issue, and the result was a corruption of ideals that was far more pervasive than the simple corruption of venality. Time after time the agency acted in fashions contrary to its own interests, and each time it appeared to learn nothing from the last time. Thus, in the Tuchman sense, it truly was an age of folly.

2

A Clear and Urgent Need

IF THERE WAS A SINGLE SERIES OF EVENTS THAT DRAMATIZED THE FOLLY OF the FDA, and transformed the modern image of the agency, it was the generic drug scandals that made the headlines and splashed across our television screens during the late 1980s. By that time the agency was perceived by many as an inefficient, indolent, uncaring, and hidebound bureaucracy; but those were adjectives that could have been equally applied to other government agencies, and to areas of the private sector as well. However, when a handful of middle-level FDA employees were caught taking cash in exchange for favors, the public view of the agency became permanently skewed. In addition to everything else, the FDA, the guardian of the nation's health, was now seen to be corrupt.

The seeds of that corruption were sown in 1984 when Congress passed the Drug Price Competition and Patent Extension Act (Hatch-Waxman), which laid down new regulations to deal with generic drugs, those copies of brand-name originals that had gone off patent. The purpose of the act was admirable in that it was designed to speed the approval process for generics through the FDA, and to make the drugs available to the public at prices substantially lower than those charged by the brand-name pharmaceutical companies, the original manufacturers and innovators.

The need to supply low-cost generics to the elderly and to those less than affluent was clear and urgent, and in its desire to accomplish that purpose the Congress crafted language into the act that limited the ability of the FDA to require proof of safety and effectiveness from generic manufacturers. The act required only that the generic maker show that the active ingredient in its product was the chemical equivalent of the original, and that it acted in the human body in the same way as the original did. Nothing more was required—the FDA was not even obliged to inspect the generic manufacturing plant before approval. The agency simply reviewed the manufacturer's Abbreviated New Drug Application (ANDA), along with the samples submitted, and either approved or denied the application on that basis.

The names of a prescription drug are manifold. Each drug has three, and all are important to the manner in which the product is marketed. The first is its chemical name, which shows its molecular structure and has an abstruse meaning only for chemists, such as 10-(3-dimethylaminopropyl)-2-chlorphenothiazine hydrochloride. The second is its public, or generic, name, which is used in the scientific literature, and is the one taught to physicians and pharmacists during their schooling. The generic name of the substance just indicated—a widely prescribed tranquilizer—is chlorpromazine hydrochloride. Finally, the drug may be known by its trademark or brand name, which the manufacturer tries to keep simple, easy to remember, and short enough to fit on a prescription blank—in this case, Thorazine. At the end of the patent protection period, usually seventeen years, the product may be marketed under its generic name, but the brand name belongs to the original manufacturer forever, except in such obvious cases as aspirin.

It is a matter of accepted business strategy that a company introducing a new, patented drug product will use the period of patent protection to mount a full-scale promotion campaign. This is aimed at not only selling the drug under its brand name while the patent lasts but also at permanently imprinting that brand name on the memories of prescribing physicians. The goal is obvious: to associate the brand name with the product so that the physician will continue to prescribe it by that name long after the patent has expired.

In most instances the results are equally obvious. Brand-named Miltown and Equanil are far more widely prescribed than meprobamate. Brand-named Lanoxin is more widely prescribed than unpatented digoxin. Serpasil is more widely prescribed than reserpine. Seconal and Nembutal are more widely prescribed than secobarbital and pentobarbital, respectively.

As an important part of this strategy, there seems to have been an effective effort to select public or generic names that are far more complex than their respective brand names, and far more difficult to remember, spell, or pronounce. Thus, most physicians find it easier to call for Kynex rather than sulfamethoxypyridazine, Daricon rather than oxyphencyclimine hydrochloride, and TAO rather than troleandomycin. Once the patent on TAO expires, anybody can market it under the name of troleandomycin—but who would want to?

Unlike the generic makers, there is no easy approval route for the brand-name manufacturers of innovator drugs. Many of these full-line companies were founded in the nineteenth century, but their exponential growth took place during the period just after World War II, along with the evolution of radically new drug therapies. Abbott, Eli Lilly, Upjohn, Parke Davis, Pfizer, Johnson & Johnson, Warner-Lambert, Lederle, Squibb, and others—then and now—invested heavily in research and development (R&D), the cost of which has been doubling every five years. In the year of this writing, the research-based pharmaceutical industry will invest $10.9 billion in R&D, which represents from 12 to 15 percent of an individual company's sales volume. No other industry spends that much on R&D in relation to sales. Agricultural chemicals, by comparison, spend about 4 percent, and electronic materials the same. And R&D is only the first hurdle in getting a new drug to the marketplace. Consider the long and expensive procedure for gaining FDA approval of the product.

Preclinical Testing. Laboratory and animal studies are done to show biological activity against the targeted disease and the compounds are evaluated for safety. These tests take approximately three and one-half years.

Investigational New Drug Application (IND). After completing preclinical testing, the company files an IND with the FDA to begin to test the drug in people. The IND becomes effective if the FDA does not disapprove it within thirty days. The IND shows results of previous experiments; how, where, and by whom the new studies will be conducted; the chemical structure of the compound; how it is thought to work in the body; any toxic effects found in the animal studies; and how the compound is manufactured. In addition, the IND must be reviewed and approved by the Institutional Review Board where the studies will be conducted, and progress reports on clinical trials must be submitted at least annually to the FDA.

Clinical Trials, Phase I. These tests take about a year and involve about twenty to eighty normal, healthy volunteers. The tests study a drug's safety profile, including the safe dosage range. The studies also determine how a drug is absorbed, distributed, metabolized, and excreted, and the duration of its action.

Clinical Trials, Phase II. In this phase, controlled studies of approximately one hundred to three hundred volunteer patients (people with the disease) assess the drug's effectiveness. This phase takes about two years.

Clinical Trials, Phase III. This phase lasts about three years and usually involves one thousand to three thousand patients in clinics and hospitals. Physicians monitor patients closely to determine efficacy and identify adverse reactions.

New Drug Application (NDA). Following the completion of all three phases of clinical trials, the company files an NDA with the FDA if the data successfully demonstrate safety and effectiveness. The NDA must contain all the scientific information that has been gathered. NDAs typically run 100,000 pages or more. By law, FDA is allowed six months to review an NDA. In almost all cases, the period between the first submission of an NDA and final FDA approval exceeds that limit; the average NDA review time for new molecular entities approved in 1991 was 30.3 months.

Expedited Process. Under a plan implemented by FDA early in 1989, Phases II and III may be combined to shave two to three years from the development process for those medicines that show sufficient promise in early testing and are targeted against serious and life-threatening diseases, such as cancer and AIDS.

Approval. Once the FDA approves an NDA, the new medicine becomes available for physicians to prescribe. The company must continue to submit periodic reports to the FDA, including any cases of adverse reactions and appropriate quality-control records. For some medicines, the FDA requires additional studies to evaluate long-term effects.

The American system of new drug approval is the most rigorous and most expensive in the world. On average, it presently takes twelve years and $231 million to get one new medicine from the laboratory bench to the pharmacy shelf, and only one in five makes it through to final approval. The generic manufacturer is spared all that time and expense, and thus is able to produce a cheap version of a safe drug to the benefit of the public.

That was the way it was supposed to work, but despite the worthy aims of Hatch-Waxman, and despite the fact that most generic drug manufacturers were honest and efficient, the act, once passed, provided an open invitation to the inept and the wicked. Not to the majority of the manufacturers. By far the largest producers of generic drugs remained the divisions of the large, old-line pharmaceutical houses that sold generics under their own brand labels. The most prominent of these in the oral tablet/capsule market were Lederle Laboratories (American Cyanamid), Warner-Chilcott (Warner-Lambert), Geneva Generics (Ciba-Geigy), Key Pharmaceuticals (Schering-Plough), and Bristol-Myers. In the injectable market, the major players were Abbott Laboratories, Baxter, Elkin-Sinn (A. H. Robins), and Solopak Labs (Smith and Nephew PLC). Between them, these old-line companies controlled an estimated 70 percent of the generics market, but the immediate aftereffect of Hatch-Waxman was the appearance of dozens of small new generic drug companies that sprang up like mushrooms after the rain.

In the year following the enactment of Hatch-Waxman the FDA was flooded with 1,069 applications for new generic drugs, compared with only 470 the previous year, and much of the increase was accounted for by the new players in the field. These were not pharmaceutical manufacturers in the traditional sense. They were not researchers or developers, some of them lacked a qualified medical director, and they also lacked the disciplines of the old-line drug companies. They were not the producers of chemicals; they bought their raw materials on the open market, and mixed them according to formula. They were cooks who were following recipes, and while the results were not always haute cuisine, they were always profitable. Dr. Marvin Seife, head of the FDA's Division of Generic Drugs, who would eventually pay an exorbitant price for playing a minor role in the scandal, said in an earlier, happier time that a generic drug factory was "a place where you put raw materials into a mixing vat, turned the spigot, and out comes gold."[1]

The gold rush was on as soon as Hatch-Waxman was signed into law, and it wasn't long before FDA inspectors were filing violation reports by the carload to the agency's headquarters in Rockville, Maryland. The law may have been made in heaven, but some of the new players weren't acting like angels. There were reports of adulterations of medicines, routine falsifications of test results, tablets contaminated with metals, and many mislabeled products.

(In one instance, a powerful psychotropic drug called haloperidol was labeled as a children's cough syrup.) Most of the offenders were caught red-handed, but anyone naive enough to believe that retributive justice would then follow swiftly was unaware of the FDA's clanking and outmoded enforcement apparatus.

From the time that an FDA field inspector discovered a problem that he felt warranted legal action, his recommendation had to clear fifteen levels of review and criticism before it reached the lawyers who handled referrals to the Department of Justice. Then there were further reviews at the Office of the General Counsel and the Justice Department before the U.S. Attorney finally got to look at the case.

And that wasn't the worst of it, for the rules of the game strictly limited the age of the evidence that the FDA could present to a court. Typically, it took headquarters in Rockville months to review the inspector's field report (some of them were two feet thick), and by the time that this procedure had been completed the evidence was often out of date. This, of course, necessitated a new inspection and new evidence, a new review and a new stack of paper, also soon out of date. After a couple of rides aboard this bureaucratic merry-go-round, the local office usually gave up and recommended a regulatory letter—a slap on the wrist that, predictably, took several more months to be issued.

The ineffectiveness of the system, as well as the inefficient and industry-oriented attitudes of the FDA reviewing officers in Rockville, became the focus in 1988 of an investigation by the House Subcommittee on Oversight and Investigations of the Committee on Energy and Commerce, whose chairman, Representative John Dingell of Michigan, has habitually monitored the agency with a ferocious intensity. Subcommittee investigators found that FDA field inspectors in Chicago had pleaded with their superiors in Rockville to take legal actions against firms that were manufacturing suspect generic drugs, and that the Rockville bureaucracy, for the most part, had rejected the field recommendations for meaningful action, and had issued a series of ineffectual warnings instead. The subcommittee found "scant evidence" that officials at Rockville were "at all concerned about the cumulative evidence that these firms were endangering the public health." Rather, the FDA reviewers "appeared disturbed that the Chicago District Office kept sending these troublesome recommendations forward."[2]

These wrongdoings uncovered by the subcommittee, although serious, were garden-variety governmental sins, no more scandalous than an Air Force purchase of a thousand-dollar toilet seat, but the scandal reached headline proportions when it was charged that at least three generic drug companies were getting speedy approval on their applications in exchange for payoffs to FDA employees.

The fraud that now began to unfold was based on the high stakes involved in being the first generic drug manufacturer to gain FDA approval on any particular product. Once the brand-name version went off patent, the first of the generics into the field could usually count on a substantial period of competition-free marketing and, despite the intentions of Hatch-Waxman, at a price set artificially high. When a widely used drug was involved, this often meant millions of dollars in profits, heady stuff for these entrepreneurial companies that had sprung up from nowhere. It wasn't just a gold rush—it was a bonanza.

"Hatch-Waxman turned the FDA's Division of Generic Drugs into a wild-west show." David W. Nelson was the chief investigator of the FDA and the pharmaceutical industry for Representative Dingell's Congressional Subcommittee on Oversight and Investigations. "It was vital for the generic manufacturer to get that first approval, and there were several ways of doing it. Some of the manufacturers paid off the FDA guys, others hired Washington law firms that had connections inside the agency and others tried through Congress, an approach that worked for some and not for others. And a lot of it was based on personality. If those people at the FDA didn't like you, you were dead, and if they did like you, it didn't make any difference what kind of crap was in your application."[3]

The exposure of the scandal began quietly enough on July 7, 1988, when Representative Dingell announced that his subcommittee had received allegations that the approval requests of certain generic drug companies were being given special treatment by FDA reviewers, while those of other companies were being hampered. Consequently, he had issued subpoenas for the books and records of several companies, which he did not name, and several people, also unnamed, including some current and former FDA employees. The action created only mild interest at first, since a similar inquiry back in 1980 had cleared five FDA employees of allegations of impropriety, but the investigations took on a more ominous tone at a subsequent hearing.

On July 28, Dingell revealed that the company involved was American Therapeutics Inc. (ATI), which had been asked to disclose any "items of value" that it had offered to FDA employees. Also named were ATI's Raju Vesegna, pharmaceutical consultant Mohammed Azeem, and FDA reviewer Charles Chang. Vesegna claimed his Fifth Amendment rights and refused to answer questions, Azeem and Chang refused to supply the requested documents, and the balloon was on the way up. Before the balloon came down again, forty-two people and ten companies had pleaded guilty to, or been convicted of, fraud or corruption charges.[4]

One of those who went to jail was Charles Chang, an FDA supervisory chemist, and according to Nelson a payoff to Chang bought not only speed for one's application but also a slowdown for the competition. "Chang had two guys who were very meticulous and who took their time with applications, and he had one guy who was so fast that he could turn out two hundred approvals a year. So if you paid off Chang, your application went to the rabbit, and at the same time your competitor's application went to one of the turtles."[5]

In order to understand the high-stakes game that Hatch-Waxman fostered, one must also understand the unusual position of the pharmaceutical industry in the United States since the end of World War II, for the industry as we know it today, research-based and international, was virtually nonexistent fifty years ago. Prior to World War II, the manufacturers of medicines were still the direct descendents of traditional chemists and druggists who were concerned with the making and selling of medicines derived from naturally occurring animal and vegetable ingredients such as belladonna, cascara, digitalis, bitter aloes, and the like. These were galenical medicines, based on the teachings of the Greek physican Galen who lived over two thousand years ago, and for the best part of those two millenia, Western pharmacology and therapeutics were based on his writings.

There had been, of course, major advances such as the discovery of chloroform and digitalis, the purification of alkaloids, and the development of vaccination; and from the fifteenth century onward there had been chemists who had started to think in terms of crude synthetic medicines. There had even been a few developments along those lines such as aspirin and salvarsan. For the most part, however, in the 1940s all that a manufacturing chemist had to offer to doctors and patients were sophisticated variations of the traditional galenical preparations. (The Eli Lilly catalogue for 1943 still listed fluid extract of dandelion.) It was only after World War II that specific active chemical ingredients began to flow out of the laboratories of the pharmaceutical industry, and to sow the seeds for a new generation of medicines.

One of the techniques that enhanced this flow was molecular manipulation, or "screening," in which slightly different forms of a basic molecule were developed in generational form to produce new drugs. Thus, in a sequence reminiscent of biblical begats, manipulation of the original sulfanilamide produced sulfapyridine, which in turn produced sulfathiazole, which led to sulfasalazine, and sulfasoxazole, and sulfamethizole. The practice of screening has been widely criticized in Congress, and by consumer advocates who view the procedure as a way of glutting the market with drugs, each with a slightly different chemical form, and each with little therapeutic advantage over existing products. The industry, however, insists that drug science has been significantly advanced by the molecular modification of existing compounds, and that these incremental gains are invariably cost effective and outnumber the relatively few new blockbuster drugs that are developed nonsequentially.

If the development of screening techniques was, and is, of questionable value to the drug-consuming public, there is no question that screening contributed in great measure to the exponential growth of the drug companies in the period following World War II. As Robert Teitelman points out, "The drug companies were not interested in science without application. Their approach was predominantly empirical, relying on what worked, without worrying overly about cause. If a compound proved effective, an organic chemist . . . would fiddle with it, snipping off a carbon atom here, adding a sugar ring there, and then screen all over again, testing all the time."[6]

The emergence of this new industry not only had an immediate scientific and technological impact on the practice of medicine but it also turned out to be a business cornucopia of unimagined riches. In the second half of the twentieth century, no major industry has been as stable and as steadily profitable as the legal drug business. Bolstered by huge cash reserves, long-term market exclusivity for its products, and the obvious, if regrettable, fact that there is no off-season for sickness, the industry is virtually recession-proof. In 1957, the annual sales of pharmaceuticals for human use stood at $1.7 billion, a decade later it had expanded to $3.0 billion, and during that same period the industry sales grew at a rate 10 percent faster than the national output of all goods and services. For the period 1967–1989, the total worldwide annual sales of members of the Pharmaceutical Manufacturers Association grew by more than 900 percent, from $4.9 billion to $50.1 billion, and during the 1980s the industry consistently ranked first in the nation in terms of profitability based on sales.

Despite such an obvious talent for turning a profit, the industry enjoys projecting an image of disinterested altruism. Profits are proudly presented at stockholders' meetings and in company reports, but to the general public the industry prefers to be thought of as a group of dedicated and ethical professionals, rather than as a band of industrial barons. In fact, the industry has a love affair with the word *ethical,* to the point where a drug that is dispensed by prescription only is known as an ethical pharmaceutical, and is not to be confused with a proprietary drug, which is purchased over the counter (OTC). Because of their link to the medical profession, the manufacturers of ethical pharmaceuticals like to think of themselves as being bound to the altruistic goals of advancing medical science, easing pain, and prolonging life. This, they insist with a certain coyness, is their primary goal, and the profits that accrue are nothing more than God's reward for diligence and good management.

While no sensible person would try to deny the enormous contributions that the pharmaceutical industry makes to our society, neither has anyone ever confused a drug maker with an eleemosynary institution. Discovering new drugs is only one aspect of the business, producing them properly is another, and selling them profitably is still a third. During the decades that followed World War II, the ethical pharmaceutical companies assured the profitability of their operations in the simplest way possible: by setting their prices at an artificially high level, and by sticking to those prices no matter what.

A case in point involving antibiotics has been provided by John Blair, who was chief economist of the Subcommittee on Antitrust and Monopoly of the U.S. Senate from 1957 to 1970. In 1951, Pfizer priced its product Terramycin at $5.10 for 16 capsules of 250 mg each, and four days later both American Cyanamid and Parke Davis announced the same price for their products Aureomycin and Chloromycetin, respectively. Two years later, American Cyanamid was the first to market the new broad-spectrum, tetracycline, setting the price at the now familiar $5.10, and shortly thereafter, in a happy coincidence, the four other sellers of that drug put their products on the market at the same price.[7]

But price fixing is one thing, and price gouging is something else. As Blair points out:

The first wide-selling corticosteroids were prednisone and prednisolone. They were followed by the introduction of methylprednisolone by Upjohn under the trade name Medrol and of triamcinolone by Squibb and Lederle under the trade names Kenacort and Aristocort. Each of these successive products [was] introduced at the price charged by Schering for Meticorten and Meticortelone, 18 cents per tablet to the druggist. This price had been established when the manufacture of corticosteroids was an expensive and complex process involving the use of oxbile, which required hundreds of slaughtered animals to yield a few grams of steroids. In its place Upjohn introduced a microbiological process, involving the use of a vegetable source, Mexican yams. In addition to replacing a scarce and costly source of supply with an abundant and inexpensive source, the new process reduced greatly the steps involved in production. . . . Yet the prices charged by Upjohn for its new corticosteroid, produced by the new low-cost microbiological process, were the same as the prices charged for the older steroids made by the high-cost oxbile method.[8]

This replacement of older high-cost products with newer low-cost products at approximately the same price was a common practice in the pharmaceutical industry, and, amazingly, nobody seemed to care. Least of all, the FDA. As Blair explains:

The attitude of a regulatory agency toward the industry under its jurisdiction undergoes a metamorphosis, changing gradually from initial hostility to a spirit of accommodation and finally to protective concern with the industry's well-being. It would be a mistake to attribute this process solely to the incompetence of government officials or to their constant departure for high-salaried positions in the companies they formerly regulated. Of at least equal importance is the constant day-to-day preoccupation with the industry and its problems.[9]

Through this process, which was termed *clientism* by Senator Paul Douglas, both the FDA and the pharmaceutical industry slowly came to believe that, in the end, they all were members of the same club.

Wallace F. Janssen, who has served with the FDA for more than forty years, recalls that in the 1960s, "Once a [pharmaceutical] firm got a license [to manufacture], they belonged to the club, and any problem had to be settled inside, in private rather than in public. All the people who belonged to the club were good people, and they would do the right thing when any problems arose. In theory, we were all ethical professionals together."[10]

After only two decades of post–World War II expansion, the leading members of the club, the dozen or so major U.S. drug companies, were the prematurely aged doyens of corporate America. They were large, they were rich, and they were very conservative. One industry analyst has pointed out that:

> Companies with patented drugs could become cash machines, particularly as manufacturing facilities were paid off and the need for marketing well-known agents declined. Without competition, with no one questioning health-care costs, they could set prices freely. And because illness paid no heed to economic cycles, they made money in good times and bad. Thus, organizationally, drug companies tended to be very stable. Few new companies could afford to enter their ranks—only Syntex had managed to do so since World War II—and management turnover was very low. Short of a recall or a scandal, top managers would spend ten or fifteen years at the top; they might be long retired before their successors discovered a bare cupboard. Even then, products going off patent faced little competition. Trying to take business from another company's off-patent drug was viewed as an ungentlemanly business.[11]

The ability of the pharmaceutical companies to set high prices during good times and bad has not faded with the years. In 1992, a study by the General Accounting Office showed that the price of prescription drugs had increased by nearly three times the rate of inflation over the previous six years. The study focused on twenty-nine common medicines, the increasing prices of which had prompted consumer complaints to the House Ways and Means Subcommittee on Health. The study showed that the prices of nineteen of those drugs had increased by more than 100 percent between 1985 and 1991, while the consumer price index for all prescription drugs increased by 67 percent during the same period, and inflation rose by only 26.2 percent. In the most flagrant example, the wholesale price paid by federal programs for 100-mg capsules of the antiseizure drug Dilantin went from $22.80 per thousand to $102.30, an increase of 348.7 percent.[12]

It was a situation that cried out for change, and in the end the changes that were effected transformed forever the manner in which pharmaceuticals would be researched, developed, and marketed. What was never changed, however, was the ability of the drug companies to set their own prices.

3

The Man in the Coonskin Cap

By the late 1950s, the ability of the pharmaceutical companies to set prices freely, and to reap enormous profits with virtually no regulation by the federal government, was a source of concern to those who would later be known as consumer advocates, to public interest groups, and to old-fashioned populists who saw the devil rising out of every industrial smokestack. Senator Estes Kefauver of Tennessee fell into none of those categories, but he was a crusading politician who had recently run out of causes. In 1950–1951, his investigations into organized crime had been nationally televised and had thrust him into the public eye. In 1952, he had sought the Democratic nomination for the presidency, losing to Adlai Stevenson. He had lost again in 1956, that time running as the vice-presidential nominee. Another losing cause of the day had been his support for the civil rights of blacks, an anomalous position for a Southern politician of that era.

Late in 1957, as chairman of the Senate Subcommittee on Antitrust and Monopoly, he called a meeting of his top staff assistants to suggest new areas for investigation, and one aide, Dr. Irene Till, urged an in-depth study of prescription drugs. Her suggestion was prompted as much by a personal animus as by an academic curiosity. Six years earlier, a doctor had prescribed an antibiotic for her husband, and the price had seemed exorbitant. Not only that, but in checking out all the competing products, she had found their prices to be exactly the same. This smacked so clearly of illegal price-fixing that she urged Kefauver to look into the area, but it wasn't until the summer of 1958 that the senator was prodded by two events into a full-scale investigation of the ethical drug business.[1]

The first prod came from the Federal Trade Commission, which for the first time that year reported the profit statements of the major drug companies as separate from the statements of the chemical industry as a whole—and the report was an eye opener. The profits of the drug companies after taxes were about 11 percent when based on sales and 19 percent when based on net worth. These were figures far above the average for nearly all other manufacturing companies and clearly could not have been attained under a normal pricing structure. The second prod came from a report published in the *Saturday Review* magazine on the false and misleading promotional campaigns that were being used by some of the major drug companies at that time. Shocked into action by these disclosures, Kefauver decided that he had found his latest cause, and subpoenas went out to the top pharmaceutical manufacturers in the country.[2]

The hearings were held off and on between December 1959 and October 1960, and at first they were taken lightly by the top management of the industry. Kefauver, by design, often affected the air of a country bumpkin in a coonskin cap, and the industry people, represented by expensive legal talent and skillful public relations manipulators, were confident that they could head off any truly damaging disclosures. They would have done well to consult with Senator Everett Dirksen of Illinois, who once had described Kefauver as having all the charm of a Victorian lady and all the single-mindedness of a hungry Apache warrior.

The first session of the subcommittee set the tone for what was to come. With Francis Brown, the president of Schering, on the

stand, Kefauver noted that Schering's product prednisolone cost
1.6 cents per tablet to produce, but was sold to the pharmacist for
17.9 cents per tablet and to the public for 28.8 cents. This
represented a markup of 1,118 percent. Kefauver also noted that
Schering marketed the hormone estradiol progynon, which it
purchased from a French firm, Roussel, and resold it at a markup of
7,079 percent.[3]

Brown's response was the classic defense that is still used by
pharmaceutical companies today: that such high prices were
necessary in order to support the company's extensive R&D
program and to aid in its mission of informing the medical
profession of the latest in pharmaceutical progress. Kefauver was
less than impressed, pointing out that Schering had done no
research whatsoever on the French drug. "You bought a finished
product from Roussel," he told Brown. "All you did was put it in a
tablet, put it out under your name, and sell it at a markup of 7,079
percent."[4]

The astronomical percentage figures made headlines the next
day, and Kefauver was back in the public eye. In session after
session, he and his aides hammered away at the drug companies,
making public the fantastic markups on one pharmaceutical
product after another, emphasizing the lack of competition in the
industry, the inordinate profits, the high cost of promotion, and the
damning fact that many of the companies were selling their
products abroad at a fraction of the prices charged to the American
consumer.

The cut-rate pricing of pharmaceuticals in markets such as Latin
America was a clear indication of the inflated prices being charged
at home, and helped to highlight the markedly different manner in
which the drug companies sold their products abroad. In the
United States the approved uses of drugs that were published in the
Physicians' Desk Reference were specific, concise, and limited to
claims that could be substantiated by scientific evidence acceptable
to the FDA. In the same sense, the warnings of possible side effects
and adverse reactions were presented in detail to the public. Not so

in a market such as Latin America where the claims for the same drug by the same company were often far more wide-ranging, and the potential hazards were glossed over or totally omitted. In some cases, only trivial side effects were described, while possible fatal reactions went unmentioned. A full decade after the Kefauver hearings, one drug company executive was able to say, unblushingly, that in the selling of oral contraceptives, "We don't say much about the need for a Mexican woman to have a regular Pap smear while she's on the Pill, because she really doesn't care. She's far less worried about the risk of cancer than she is about having her umpteenth unwanted baby."[5]

In their own defense, the drug companies pointed out that such practices did not violate any of the local laws of the various countries involved, but hawking potentially dangerous drugs to a population that was both undereducated and naively trusting was light-years away from the altruistic image that the industry preferred to project.

Throughout most of 1959, the off-again on-again hearings of Kefauver's subcommittee kept the American consuming public focused on the least attractive faces of the pharmaceutical business as the senator and his aides ripped away at the altruistic facade that the companies had worked so hard to erect, fostering a climate that demanded an increased regulatory control over the industry. The sort of control that Kefauver had in mind was limited, at first, to the high cost of pharmaceuticals, something that could only be regulated by the FDA, and the senator set out to secure the cooperation of the agency in backing the necessary legislation.

Winton B. Rankin, who eventually retired from the FDA as deputy commissioner of Food and Drugs, remembers that Kefauver called him and asked if he could come over and talk about the situation with the then-Commissioner George Larrick. That, in itself, was an unusual request, and it showed the Kefauver touch. "Ordinarily a senator phoned down or had his staff phone down and say, 'You come up and talk with me,' [but] Senator Kefauver and John Blair, his staff economist, and two or three other people came down to visit Commissioner Larrick in his office. What it amounted to was that Senator Kefauver wanted Commissioner Larrick to support him in the introduction of legislation to deal with the economics of the drug industry."[6]

Larrick, who headed the agency for eleven years, was the last of the homegrown FDA commissioners, career men who had risen through the levels of the bureaucracy to the top job. Beginning in 1965, when Larrick's successor, Dr. James Goddard, was brought in from the Centers for Disease Control to shake up the agency, the job of commissioner became a political appointment, and it has stayed that way. But Larrick was FDA all the way, with traditional FDA concepts about the role of the agency and its relationship with the pharmaceutical industry, and the last area in which he wanted to be involved was the high cost of drugs. That, according to the code of the club, was none of his business, but rather something for the ethical professionals to sort out among themselves, and in private.

Rankin recalls that the commissioner was polite but noncommittal to the senator and his aides. "Irene Till and John Blair were pressing ahead just as hard as they could on the economic aspects of drug regulation, but Commissioner Larrick indicated that he did not believe that this was an area for the FDA, that our area was health, not economics."[7]

The area that the FDA was concerned about was the efficacy of the drugs that it approved for the marketplace. At that time, a manufacturer seeking FDA approval for a new drug needed only to prove that the product was safe to use, and without any hazardous side effects. What the agency wanted was legislation that would compel the manufacturer to prove that the drug was efficacious, that it did what it was supposed to do, that it actually worked.

"At that moment," recalls Rankin, "Larrick didn't want to have a darn thing to do with Kefauver's bill. He saw it as just a lot of trouble for Food and Drug that was not going to help the agency."[8]

A possible explanation for this foot-dragging by top FDA officials lies in the unhappy story of Dr. Henry Welch, who had been head of the agency's antibiotic division. At the same time, he had been editor in chief of two journals, *Antibiotics and Chemotherapy* and *Antibiotic Medicine and Clinical Therapy,* and also co-owner of the *Medical Encylopedia.* This clear and questionable case of academic moonlighting had come under the scrutiny of Kefauver's subcommittee, and a deep probe into Welch's activities had uncovered the fact that he was also working in undue closeness with such pharmaceutical firms as Parke Davis and Pfizer. Welch admitted to government investigators that he did, indeed, receive "a small honorarium" for his services, but when the total of those payments turned out to be in excess of a quarter of a million dollars, the good doctor decided to take early retirement from the FDA.[9]

The top echelon at FDA never forgave Kefauver for what it considered to be the hounding of Welch from public life, an attitude indicative of the agency's mindset whenever it comes under outside criticism. The wagons are circled and the attackers are fought to the hilt, no matter what the merits of the case. It is a mentality that admits to wrongdoing only under the most egregious of circumstances, and one that never, ever, forgets who pointed the finger. Welch's colleagues may have been unhappy about his revealed activities—they may even have been shocked—but they reserved their anger for "those people on the Hill" who had blackened the name of one of their own. After the Welch episode, Kefauver was a senator who could be used to help the agency obtain its goals, someone to be manipulated, but never to be fully trusted.

Even if the FDA attitude toward Kefauver had not been clouded with suspicion, there seemed to be little incentive to work with him. Kefauver was concerned with pricing, the agency was concerned with efficacy, and as Rankin recalls, "Larrick didn't believe then, and none of the rest of us believed, that there was any possibility in the next few years of getting an efficacy provision in the drug chapter of the law."[10]

Their belief was well founded in the realistic understanding that the drug industry would resist efficacy testing with every resource at its command. Testing for safety was expensive and time-consuming enough, but an approval process that mandated testing for efficacy as well promised an exponential increase in both time and money, a promise that we have seen kept. Anyone familiar with the workings of the drug companies, with their massive resources, and with their willingness to dispute any attempt at progressive legislation had serious doubts that an efficacy provision could ever be put into place. The industry simply had too much muscle and too much money, both commodities easily translatable into legislative clout. Given those circumstances, the odds were great indeed against the passage of any form of efficacy amendment to the basic Food, Drug, and Cosmetics Act, and those odds were beaten only by a serendipitous combination of political realities and a medical tragedy that was only narrowly avoided.

Beginning in 1959, a marked increase in the number of infants born with a severe deformity of the limbs called phocomelia was noted in a number of European countries, and eventually was linked to the use of the tranquilizer thalidomide during pregnancy. Mothers who took the drug during the first trimester, when the limb buds of the fetus are formed, produced children with a wide but distinctive range of deformities. Some had no arms, just flippers extending from the shoulders; others had limbless trunks with toes extending from their hips; others were born with just a head and a torso. It was the nightmare come true of every prospective parent. Thalidomide was manufactured by Chemie Grunenthal of Germany, and was sold over the counter and by prescription by many firms in many countries under licenses from the parent company. In the end, more than eight thousand children in forty-six countries were affected. The American company that applied to the FDA to market the drug under the brand name Kevodon was Richardson-Merrell.

Thalidomide was not a case of a good drug that went bad, for there was no clear-cut therapeutic use for the compound. It was not a lifesaving drug, but only one of many tranquilizers that had come onto the market in the decade after World War II. It was promoted by its maker as being nontoxic, with no side effects, and completely safe for pregnant women. Not one of those statements was true. Even if thalidomide had not been capable of teratogenic activity—an effect on the fetus quite different from the effect on the mother—it should not have been allowed on the market because it produced peripheral neuritis as a side effect, a numbing of the hands and feet that is irreversible.

There were many scientific tests that, had they been made, would have shown thalidomide to be both unsafe and ineffective, but the drug companies involved simply did not perform them. They were not required to by law.

That thalidomide was never marketed in the United States was largely due to the stubborn skepticism of the FDA's Dr. Frances Kelsey, whose doubts about the drug kept it out of American pharmacies. She fought a dogged defensive battle, blocking and parrying every attempt by Richardson-Merrell to gain approval until the news from the European countries made approval unthinkable. In the aftermath of the thalidomide revelations the drug companies, in an orgy of defensive criticism, would repeatedly raise the question: Did Dr. Kelsey get thalidomide right for the wrong reasons? Did she behave merely as a cautious bureaucrat in delaying approval of the drug time after time? Getting something as momentous as thalidomide right for the wrong reason is a perfectly respectable position, but a look at the record shows that Kelsey got it right largely for the right reasons, indeed. It is true that she did not predict that thalidomide would cause birth deformities, but she did raise the question.

Dr. Kelsey was particularly interested in fetal damage because during the 1940s she had worked on the antimalarial drug quinine, which had been found to possess teratogenic activity, but her concern was more fundamental than that. She wanted to know how the drug behaved in the human body. She wanted to know about its stability, about its effect on human metabolism, about its basic chemistry and pharmacology. She did not want to know how the drug worked in laboratory animals; she wanted to know how it worked in humans. She wanted to know the answers to questions that often were not asked in those days because no law required them to be asked. Neither Chemie Grunenthal nor Richardson-Merrell, the American licensee, could or would answer her questions, and so the drug went unapproved in the United States.

When the thalidomide story finally broke, the newspaper photographs and television clips of armless and legless European children had a shocking effect on the American public, evoking a nationwide murmuring of "there but for the grace of God. . . . " The American people had been slapped in the face with the manifest need for new drug regulations, and when it was revealed that intense industry pressure to approve thalidomide had been brought on both Kelsey and the agency, the legislative clout of the drug companies began to dissipate rapidly. No amount of money and muscle could have stood up against those photographs. "Estes Kefauver's bill for the control of the marketing of drugs lingered in committee to immense public indifference until the thalidomide scandal provoked national anger and congressional action."[11]

Kefauver now had a clear shot at the passage of a regulatory amendment, but the nature of that amendment was in doubt. The senator was still focused on what he considered to be the obscenely high cost of certain pharmaceuticals, and the need to make drugs more affordable to the average family, but the thalidomide story had shifted the national attention to the areas of safety and efficacy, and at that point Kefauver began to feel pressure from the Kennedy White House. The two men had been senators together, but the relationship had been strained since the Democratic National Convention of 1956 when John F. Kennedy had contested the nomination for vice-president and had lost it to Kefauver. Kennedy enmity tended to fade slowly, and he was not in the business of giving away political poker chips.

Rankin recalls that, "The administration didn't want Mr. Kefauver to get credit for an important piece of drug legislation, and the directive came down to our department from the White

House to come up with an alternative bill that we could call the administration bill. Actually, they didn't want a bill at that time, but when it became apparent that Kefauver was apt to get something, then they wanted an administration bill that was divorced from the economics provisions of the Kefauver legislation."[12]

It has never been fully explained why Kennedy, like the FDA, was not inclined to interfere with the pricing structure of the pharmaceutical industry. His biographers do not consider it a matter of moment. Perhaps it arose from his strained relations with Kefauver, and a desire to distance the administration from the senator's bill; and perhaps it arose from the sincere belief that efficacy was more important than economics. That his position reflected a sympathetic attitude toward drug company profits is most unlikely when attached to the man who once was quoted as saying, "My father always told me that all businessmen were sons of bitches, but I never really believed him until now."[13]

Whatever Kennedy's reasons, the political lines were now defined, and the horse-trading phase of the legislative process took over. Kefauver had one set of priorities, the FDA another, the White House a third; and the drug companies had their own defensive strategy, which was to block every attempt to control the price of drugs. The companies knew that they were going to be stuck with some form of efficacy legislation. Thalidomide had seen to that. The companies faced a future that would burden them with a costly and extensive approval process, but no matter how costly and extensive, those expenses could always be passed on to the consumer so long as the companies were free to set the price. They conducted their campaign on solid terrain. The FDA wanted nothing to do with controlling prices, nor did the White House. The concept had been Kefauver's alone, and now even his position on prices began to erode. Because of thalidomide, the efficacy and safety of prescription drugs were the flags that were flying the highest, and Estes Kefauver knew when it was time to salute. As Rankin remembers it, "Mr. Kefauver grabbed [the efficacy issue] and began running with it."[14]

Kefauver never stopped running, and he ran so fast that he left the issue of pricing far behind. His amendments required that before a new drug could be approved by the FDA, it had to be proven both safe and effective. Advertising and labeling of every drug had to include both the brand name and the generic name, plus warnings of contraindications and potential side effects. Drug plants had to be regularly inspected by the FDA, and the agency had to be given ample time to evaluate a new drug before approval.

The Kefauver-Harris Drug Amendments to the Food, Drug, and Cosmetics Act were a landmark piece of legislation that forever changed the functions of the FDA, its relationship with the pharmaceutical industry, and its responsibilities to the nation. Kefauver-Harris produced far-reaching changes in the ways that ethical pharmaceuticals were created, developed, and sold to the public, but, ironically, not one of those far-reaching changes had anything to do with controlling the high cost of prescription drugs. The issue was simply traded away, and it would be twenty years before the problem would be squarely faced by the passage of the Hatch-Waxman Act.

4

A Mess in Rockville

As Cato the Elder was to the Roman Senate, and Pitt the Elder was to the House of Commons, so is John Dingell the Younger to the U.S. House of Representatives. A congressman for thirty-five years and the son of a congressman (he occupies his father's seat), the Democrat from Michigan has acquired the reputation of being the most feared man on Capitol Hill, a tireless critic, a gadfly, and a tormentor of those whom he perceives to be either evil or inefficient. From his position as chairman of the House Energy and Commerce Committee, and its powerful Subcommittee on Oversight and Investigations, Dingell has gone after half the agencies in Washington, and during the 1980s he helped either to embarrass or to bring down such Reagan administration figures as top White House aide Michael Deaver, Interior Secretary James Watt, and Environmental Protection Administrator Ann Gorsuch. When Reagan's Secretary of Transportation Elizabeth Dole wanted to sell the government-owned Conrail to Norfolk Southern railroad, Dingell objected, and proposed a public offering instead. Dingell prevailed.

Dingell prevails more often than not, and his interests range from environmental and health issues to the telecommunications

and securities industries. As a result, the domineering congressman
has accumulated a variety of nicknames, ranging from "Big John,"
to "The Truck," to "The Samurai Warrior from Michigan," and to
"Tailpipe Johnny," in acknowledgment of his loyalty to the Detroit
motor industry.

But despite his wide range of interests, Dingell seems to save his
heaviest fire for the FDA, bombarding the agency with hearings and
investigations on such diverse subjects as seafood inspection, the
safety of the nation's blood supply, the regulation of medical
devices, and the purity of imported foods. Notification of a Dingell
hearing sends almost palpable tremors through the FDA
headquarters building at Parklawn, and lights burn late as officials
at the highest level prepare themselves for a grilling that they know
will be both merciless and tireless.

Such was the case in May 1989 when the Subcommittee on
Oversight and Investigations held hearings to look into allegations
brought by three manufacturers of generic drugs. The three—Barr
Laboratories, Mylan Laboratories, and Barre-National—all made
substantially the same accusations: that certain of their competitors
had been bribing FDA reviewers in order to gain preferential
treatment for their applications. Further, they contended that when
their complaints were made to the highest levels at the FDA,
including the director of the Division of Generic Drugs, Dr. Marvin
Seife, they were treated with indifference at best, and abuse at worst.

Seife, himself, testified voluntarily on May 3, 1989. Amiable and
well liked, he had, according to colleagues, a reputation as a
straight shooter. "He was charismatic," said one, "the kind of
bureaucrat that made you proud of working for the government—
and he demanded an honest day's work. He was in at 6:30 in the
morning, and he was ready to go. That's the way he liked to run
the shop." With Seife at its head, the Generic Drugs Division was
given "lots of awards" for its work in implementing the Hatch-
Waxman Act, and as late as 1986, Seife was given the FDA's Award
of Merit for "consistent and sustained excellence of performance in
ensuring drug safety and efficacy."[1]

And his reputation as a straight shooter was apparently well earned. Under questioning by the subcommittee's investigator, David Nelson, Seife related that in 1974 he was on the receiving end of an unsolicited gift, a Panasonic color television set, from Bolar Pharmaceutical of Copiague, New York. "I was shocked and chagrined," he told the subcommittee. "I repacked the container . . . and returned it to the firm." Bolar promptly apologized for this breach of etiquette.[2]

Seife's personal probity was never in question, but his testimony before the subcommittee indicated that he was considerably less than a disciplinarian. In fact, he ran so loose a ship that corruption was able to flourish without restraint, and he finally took action only when faced with unalterable facts. He also had one simple pleasure that eventually would cost him dearly. He enjoyed having companionable, and strictly forbidden, lunches with members of the pharmaceutical trade. He enjoyed them so much and so flagrantly that nine years before the hearings he had been temporarily suspended from his job for that sort of indulgence. Suspended at the same time, and for the same offense, had been the ubiquitous Charles Chang, and several others. As Seife quaintly put it, "This removel of Mr. Meyer, myself and some ladies, and Mr. Chang occurred in March 1980."[3]

David Nelson, questioning for the subcommittee, pointed out that at the time there had been allegations made about a number of lunches, some liquor, and some fur coats.

Seife replied, "I never—I had heard some of this material is rumor, I had never—nobody in the Agency ever told me that this was a fact. I'd heard rumors about Mr. Chang. I knew that one of the ladies had a little live lobster business that she was pursuing. I don't know about gifts to the others. In my case, it had to do with occasional lunches with members of the trade."[4]

Under questioning, Seife went on to tell the subcommitte of how, in December 1985, he had received an anonymous telephone call from a man who "apparently had left the employment of . . . American Therapeutics." The caller told Seife that American

Therapeutics had been in the practice of supplying a girlfriend for Chang, who had been spending weekends with the unknown lady. Seife testified that he had relayed this information to an FDA conflict of interest group, but that he did not know if any investigation was made.[5]

Seife did nothing to discipline Chang at that point, but in January 1986, according to his testimony, he began to become suspicious of various applicants. "The first one happened to be American Therapeutics. [I decided that] all new applications coming into the Division from American Therapeutics were [to be] given to other review branches [not to Chang]. This was followed by a number of other companies as my heightened suspicion grew."[6]

Seife's further testimony presented a picture of the complex interrelationships within the generic drug trade, a picture of greedy drug companies scrambling for position in an explosive market situation, and equally greedy FDA examiners who were willing to accommodate them. There were tales of a fur coat and a case of wine, of free transportation,[7] and of an enigmatic drug firm consultant named Mohammed Azeem who had developed the disturbing habit of dropping envelopes stuffed with thousands of dollars in cash onto the desks of FDA examiners.[8]

There was My-K Laboratories and its principal, K. C. Bae. Seife testified that, "Mr. Bae was trotted out by the previous [Reagan] administration as their minority showpiece. He apparently got a Small Business Administration loan and was put into—helped into business—and Governor Thompson used him as did President Reagan."[9]

Seife's story went on for six hours in closed session. His personal honesty never in question, he provided what Representative Dingell would later call a "road map" of suspected corruption in the generic industry and the FDA that would later lead to a slew of fines and jail sentences for drug company executives and FDA employees.[10]

That should have been the end of it for Marvin Seife. If he had been something less than severe with his corrupt employees, and somewhat less than prompt in coming to realize the seriousness of the situation, he had made the right moves in the end, and his voluntary testimony before the subcommittee had spread some much-needed sunshine on the situation.

But that wasn't the end of it. There were still some lunches to be accounted for. As the *Washington Post* later put it, there was just something about Marvin Seife and lunch. In the late 1970s, it was common practice for drug company representatives to fly into Washington and lunch with Seife three or four times a week. This was against regulations and, as we have seen, Seife was officially reprimanded for the practice in 1980, and he signed a letter in which he promised to change his ways. Nine years later, in the wake of the subcommittee hearings, he signed an affidavit stating that it was his policy "never" to have such lunches. This was a lie, and the affidavit was his undoing. Charged with perjury, Seife was brought to trial in October 1990, and the government was able to document a dozen lunches that he had had with industry representatives over a three-year period.[11]

That was the first count in the indictment. The second count involved a particular lunch with K. C. Bae, which was also attended by a representative from Warner-Lambert. Bae later pleaded guilty to a racketeering charge in connection with a bribe paid to one of Seife's subordinates, and Seife, in his written affidavit, had specifically denied that he had ever had lunch with either Bae or the drug company rep.[12]

The charges brought against Marvin Seife are difficult to take seriously when placed against the background of the entire generic drug scandal. The man was not corrupted by fast women and slow horses. No one bought him a Bentley, or a case of champagne. He simply enjoyed a companionable lunch once in a while. Lunch was the high point of his day, and what better way to spend it than in the company of his friends and colleagues in industry? But the government did not see it that way. The U.S. attorney who prosecuted the case, Breckinridge L. Willcox, accused Seife of setting a bad example, which tended to encourage some of his subordinates to engage in far more venal activities.[13]

With a distinct air of sanctimony, Willcox said at the time that, "We're not in the habit of indicting government employees for taking meals, but it was the lying about it [that went] beyond the pale. He [Seife] set the moral tone for this absolutely corrupt office that he was in charge of."[14]

Other observers close to the investigation were not nearly so righteous. "Nobody thought that [Seife] was a crook in the way that the manufacturers or those who took bribes were crooks," said one of them. "He was a complicated man who did a lot of good . . . but he had a blind spot on lunches . . . and [he] made the mistake of lying about it."[15]

Seife was transferred out of the Generic Drug Division in 1989, and at the end of that year he retired to San Antonio, Texas. The indictment came the following year, and he was convicted and sentenced to five months in a work-release facility, with another five months of home detention. But federal sentencing guidelines required actual prison time to be served, and on February 10, 1992, Seife turned himself in at the federal prison in Three Rivers, Texas, to begin doing his time. It was at this point that Marvin Seife's life began to unravel in earnest.

Seife and his wife, Norma, arrived at Three Rivers at 10:30 at night, and were told that his paperwork was missing. As John M. Oury, the associate warden at the facility later explained, when someone surrenders without official papers at the Federal Bureau of Prisons, "we have only their word, and our practice is to button that person down until we know what we have in hand."[16]

The bureau's method for buttoning Seife down was to throw him into solitary confinement for twelve days while they waited for his paperwork to arrive. Included in that paperwork was a letter from his personal physician stating that Seife was prone to serious infections. While he was in solitary, the authorities confiscated his shoes, left him to walk around in socks, and finally issued boots that were too small and blistered his feet. The blisters became infected, the infection spread, and Seife nearly died before he was driven 40 miles to the Bee County Regional Medical Center, where a surgeon amputated his gangrenous left leg below the knee.

This was all in the future on May 11, 1989, when the three complaining generic drug manufacturers appeared before the Dingell Subcommittee to make their charges of bribery and preferential treatment. The first company witness, Roy McKnight, chief executive officer of Mylan, said that he was appearing with a mixture of pride, anger, and relief. He said that he was proud of what the generic drug industry had accomplished for American consumers, but that he was angered by the mounting evidence of lawlessness and corruption in the Generic Drug Division of the FDA. He claimed that FDA personnel repeatedly refused to meet with Mylan representatives who were regularly subjected to public abuse, and he was especially incensed at Dr. Seife who, he claimed, "steadfastly refused to see us and became increasingly hostile."[17]

McKnight charged that because of Mylan's complaints about FDA corruption, a long period of open animosity on the part of the agency had ensued, during which the FDA's "unwarranted actions were extremely damaging to our market position." He cited examples, one of which was clonidine, a blood pressure drug that was a generic version of Boehringer-Ingelheim's drug Catapres. According to McKnight, when the application for clonidine seemed just about ready for approval, it was inexplicably reassigned to a second reviewer, who was instructed to begin the process all over again. When the second reviewer recommended immediate approval, Charles Chang, her supervisor, ordered her to let the application sit on the shelf for two months, then reassigned it to a third reviewer who began the process all over again.

McKnight also charged that Mylan's application for clorazepate (a generic version of Abbott's Tranxene), an antianxiety drug, was delayed so that "a brand new untested company, Able Laboratories, would have earlier access to the market." Furthermore, when Mylan got approval for a biostudy on ibuprofen (Upjohn's Motrin), the FDA decided to institute new rules for the study, but did not notify Mylan of the change—unlike its competitors—for six months. McKnight estimated that this oversight cost the company $10 million in sales.[18]

In the spring of 1989, Mylan became fully convinced of FDA discrimination, and it confirmed a rumor that Chang had gone so far as to instruct his reviewers to lose jackets, slow down reviews, and fabricate excuses to prevent Mylan from receiving approvals. McKnight told the subcommittee:

> We were forced to face a stark reality. We decided it was time to launch an investigation of our own. We knew it was a long shot. It was well known throughout the industry that so-called complainers could expect to receive the harshest of treatment, including direct retaliation. After a year of investigation, it was apparent that the problem was not simply Mr. Chang, but a far broader pattern of improprieties by many companies doing business with the Generic Division. For a time we were stunned. We had grave doubts about the willingness of the FDA or any agency within the Department of Health and Human Services to thoroughly probe the situation.[19]

McKnight said that Mylan felt that it was "at the mercy" of the FDA, since Dr. Seife refused to take action against Chang, and appeals to Seife's superiors proved to be fruitless. It was at that point that Mylan began its own investigation, which ultimately uncovered substantial evidence of Chang's wrongdoing.[20]

"We decided," he said, "to bring our information to the attention of this Subcommittee. That was eleven months ago."[21]

McKnight further testified that the FDA response to these charges had been threats of reprisal "neatly planted in the trade press suggesting that Mylan was foolish to bring its complaints to the Subcommittee, and promising that people in the agency . . . have long memories."[22]

Max Mendelsohn, the president of Barre-National, presented similar testimony, and then it was the turn of Edwin Cohen, president of Barr Laboratories. Cohen, feisty and outspoken, presented additional evidence of arbitrary and unfair actions by the

FDA concerning generic drugs. These included retroactive decision making, shifting standards, procedural and substantive leaks, favoritism by a reviewer, and high-handed and arrogant treatment by FDA officials.

But the FDA had had enough. Wounded in pride, dismayed by the revelations that were coming out before the subcommittee, and deeply in need of self-justification, the agency struck back in an unparalleled display of institutional folly. One hour after Cohen had finished testifying, FDA inspectors swarmed over Barr's facilities in Pomona, New York, and Northvale, New Jersey, in a surprise inspection for manufacturing violations. The inspection was called off two hours later, but it was clear that this was only the opening gun in a campaign of calculated harassment. Given its historical posture of aggressive defensiveness when accused of even the most casual of sins, the agency reacted predictably. In the series of lawsuits between Barr and the government that followed, Barr asked the courts for relief from harassment, and the FDA, quite candidly, stated that it wanted to close down Barr. The FDA never forgives a whistle-blower, regardless of the rights and wrongs of the case.

Industry analysts generally agreed that the government seemed set on showing that it would not tolerate challenges from drug companies. "It's like a policeman giving you a speeding ticket," observed Hemant K. Shah, an independent analyst in Warren, New Jersey. "You don't punch him in the nose."[23]

According to Barr, for the next three years the FDA never stopped trying to put the company out of business, peppering it with minor inspection violations and in some cases substantially delaying approval of Barr's applications for generic products. One of these was a generic version of erythromycin estolate, an antibiotic, which Barr had submitted for approval in January 1987. The approval wasn't granted until October 1990, an unusually long delay, which Cohen called a reprisal for his accusations before the Dingell Subcommittee, and which the FDA attributed only to a shortage of employees.

In another case, one with more far-reaching ramifications, Barr brought suit against the FDA, alleging that the agency had leaked proprietary information contained in Barr's application to manufacture a generic version of conjugated estrogens, a menopausal drug. Barr contended that an FDA reviewer passed along information from the application to a representative of Wyeth-Ayerst, the manufacturer that originally had developed the drug, and that, as a result, millions of American women had been denied economically feasible access to a drug that eased the effects of menopause and lessened the chances of heart attacks for postmenopausal women.[24]

Barr's contention was based on the fact that there are about 35 million women in America who are of postmenopausal age, and that almost every one of them should be on hormone replacement therapy—not only because such therapy eases the well-known symptoms of menopause such as hot flashes, and not only because it aids in protecting bone mass, but also because of its effect on the cardiovascular system. Cardiovascular disease is the major killer of women in the United States, eight times more likely to kill than breast cancer. But hormone replacement therapy was being denied to many women because of its cost, and because there was no generic version of conjugated estrogens on the market. Only Premarin, the original product manufactured by Wyeth-Ayerst, was available, but at a price much higher than a generic version would cost, one much too high for women with limited incomes. And that cost was increasing every year. According to a report issued by the Government Accounting Office in August 1992, the cost of Premarin had increased by 160 percent over a six-year period ending December 31, 1991.[25]

The appearance of a generic drug on the market does not always result in the lowering of the name-brand price, despite the competition. Often the innovator company actually raises its price to compensate for lost sales, while counting on customer loyalty to the brand name. But in the opinion of industry analysts, a generic competitor to Premarin would have resulted in an overall drop in the price of conjugated estrogens. Barr's generic version of the drug had been virtually assured of FDA approval, and the company insisted that the leaking of confidential information to Wyeth-Ayerst resulted in the agency abruptly rescinding its guidelines on the drug's testing methods, thereby voiding Barr's application. The FDA, which had agreed to accept clinical testing based on urinary metabolites, suddenly decreed that the testing would have to be based on blood plasmas. Aside from being of questionable scientific validity, the FDA shift nullified years of research, cost Barr an estimated $2.5 million, and delayed the marketing of generic conjugated estrogens by at least two years. The delay continues to the day of this writing.[26]

The FDA reacted to the charges of harassment and retaliation in the time-honored manner of all bureaucracies. In September 1990, Acting Commissioner James Benson appointed a three-man panel to investigate and respond to the charges. The appointees were announced as being unaffiliated with the FDA, but they were, in fact, far from neutral. The first was the inspector general of HHS, the parent body of FDA; the second was a senior executive of the American Pharmaceutical Association; and the third was a private consultant who previously had been employed by a major pharmaceutical company. Even this panel, which hardly could be thought of as disinterested, had difficulty bringing in a unanimous decision, and in the end each of the three filed his own report. Not surprisingly, each decided that Barr's difficulties with the FDA did not arise from agency retaliation, but one concluded that, "Barr was well known to the agency and inclusion [harassment] might have been welcomed by some," and another found that, "The denial of Barr's ANDA for conjugated estrogens was neither fair nor equitable." The third member agreed that Barr had been compromised regarding its conjugated estrogen application, and reported that, "Special and/or preferred treatment appears to have been a true situation when Dr. Seife was in charge."[27]

Despite its denial of agency retaliation, the combined reports have a ring both of truth and of common sense. It is difficult to imagine a calculated decision at the highest levels of FDA to "get" Barr. But it is quite easy to imagine the unofficial reactions at lower levels to a manufacturer who not only had blown the whistle on friends and colleagues but had challenged the primacy of FDA in pharmaceutical affairs. It is quite likely that no generals made the decision, but rather the sergeants of the FDA, the middle-level bureaucrats to whom face is everything. For these are people who know that they are virtually futureless, that they have gone as far as they ever will. Unlike their counterparts in industry, no lucky stroke or sudden breakthrough is going to raise them out of the heap. All that lies ahead is, perhaps, one more step up the GS (government service) ladder, one more minuscule pay raise soon to be swallowed by an uncompromising rate of inflation.

People in such a position are sustained by the twin crutches of power and respect. Power provides the muscle flex that almost— but not quite—makes up for money, while respect provides the mask for mediocrity. Challenge either, and you challenge the soul of the bureaucrat.

Barr had flung down the challenge, and the bureaucracy had reacted in the only way it knew. No one had to give an actual order. For Henry II, it was enough to wonder rhetorically if no one would rid him of a meddlesome priest, and Thomas Beckett was as good as dead. In a bureaucratic equivalent of a murder in the cathedral, someone had to get Barr. Besides, went the common wisdom, Barr was probably guilty of something. Everybody was guilty of something.

This common wisdom often prevails among those charged with enforcing a law or a regulation. Jerome Greenfield, writing about an earlier and quite different case at FDA in his book, *Wilhelm Reich vs. the USA*, remarks that he "spoke with FDA officials about the preliminary investigation and the fact that it was conducted not to discover the truth but, rather, proceeded on an a priori assumption of guilt." One high official in the Office of Compliance, who had not been personally involved but was acquainted with the case, said: "We're cops and that's part of the cop psychology.

When an inspector is told to investigate some operation, he automatically proceeds on the basis of there being something illegal involved that he's got to find out about."[28]

It is a sadness of language that a cliché gets to be what it is through the confirmation of experience. Nine times out of ten you cannot fight City Hall, and ten times out of ten you do not punch the traffic cop in the nose. Thus it is in matters of regulatory compliance. In cases such as these, the FDA always wins. There are other instances in which FDA procedure can be, and has been, controverted by public opinion. One such example was the outcry from the AIDS community over the approval process for new drugs.

But in matters of compliance, the FDA is supreme. If the agency says that the floor of your plant is dirty, then you'd better grab a mop and a broom, even if you know that it's pristine clean, because no court in the country will take your side. This is something that Barr learned the hard way. Three years and four lawsuits later, after watching its shares plummet from a twelve-month high of $37.50 to $6 in the middle of 1992, after stopping production of several lines of products and reducing its work force by 25 percent, Barr finally called it quits. Early in 1993, Edwin Cohen was phased out of his power position in the company, and Washington lawyer Bruce Downey took over as CEO with the avowed purpose of making peace with the FDA. City Hall had won again.[29]

In the wake of the scandal, the generic drug industry promised to reform itself, and the FDA instituted stiffer inspection and review practices. But left in question were the many generics that had been approved during the suspect period of 1984–1989, particularly a group of twenty-four drugs that are said to operate in the narrow therapeutic range.

These are drugs that are prescribed for conditions such as hypertension, stress, heart arrhythmia, and asthma—drugs in which there is a relatively small difference between a helpful dose and a harmful one. Not so with drugs that have a broad therapeutic range. Think, for example, of cases in which the same dosage of an antibiotic will be prescribed for a 125-pound woman and a 200-pound man. Both bodies will absorb a dosage well within the broad therapeutic range without harm, but the situation is quite different for the narrow-range drugs.

Deviations of more than 10 percent between generic and brand-name products are rare; usually the ranges are much less than 10 percent, but with narrow-range drugs even a small variation can be cause for concern if the dosage climbs to toxic levels. Toxicity from the narrow-range antiasthmatic drug aminophylline, for example, can result in nausea, vomiting, and convulsions, while an overdose of oral diopyramide, used to treat irregular heartbeats, can lead to loss of consciousness and death. Overdosing any narrow-range drug, brand-name or generic, is carefully avoided by physicians, but the generics raised a particular question after the scandal because, in the opinion of some experts, their dosages had not been satisfactorily established. The question was raised because of the methods used by the FDA in retesting narrow-range therapeutic drugs after the scandal broke. The need for retesting was clear, but the FDA chose a method that was sure to arouse controversy. The agency chose to retest in vitro, not in vivo.

The difference between the two methods bears explanation, since the name of the game in such testing is *bioequivalency,* the need to show that the generic drug works in the body in the same fashion as does the original. During the course of the development of an original drug, the testing for safety and dosage is done in vivo, in the living human body over a long period of time. But the generic manufacturer, who is simply copying the original formulation, is under no obligation to perform such extensive testing, and instead uses dissolution studies in vitro, in laboratory vessels filled with liquids. The difference between in vivo and in vitro is sharp and clear. Dissolution testing, in vitro, is more a quality-control method than an actual test. It suggests how rapidly the product will dissolve in a simulated fluid, and is used to assure lot-to-lot similarity when a product is being manufactured over and over again. The one form of test can never be a substitute for the other, and as a congressional critic of the FDA put it derisively, "Dissolution testing is like dropping a pill in a bucket of water."[30] The point was well established in 1988 by the FDA, itself, when one of its task forces reported that, "There is not yet evidence to show that any particular dissolution pattern [in vitro] alone will assure bioequivalence [in vivo]."[31]

But if there was no such evidence, why was the FDA content to retest in vitro? The official FDA response, that an in vivo study would have been time-consuming and expensive, was both predictable and unfortunate. These were not ordinary products— these were extraordinarily sensitive drugs—but they were treated in a routine fashion. Considering the circumstances, nothing should have been spared in terms of time and expense to retest these drugs, and the question must arise as to how much incentive there was at FDA at the time to confirm its own culpability. Whatever the reason, those twenty-four drugs were retested and confirmed as being bioequivalent with the innovator drugs under conditions that were considerably less than ideal.

When the dust from the generic drug scandal finally settled, forty-two people and ten companies had pleaded guilty to, or been convicted of, fraud or corruption charges. The stiffest sentence went to Robert Shulman, the former president of Bolar Pharmaceutical Company, who was given five years and fined $1.25 million.[32] The body count was impressive, but not all-inclusive, for there were intangibles that could not be calculated.

Beyond calculation was the damage done to the generic drug industry, for the most part composed of honest businessmen and businesswomen involved in providing a necessary product at a price that most people could afford.

Beyond calculation was the damage done to the morale of the men and women of the Generic Drug Division, and the FDA as a whole.

Beyond calculation was the damage done to firms such as Barr and Mylan, which had blown the whistle, pointed the finger, and suffered the wrath of the FDA.

Beyond calculation was the damage done to millions of postmenopausal women who were denied affordable access to hormone replacement therapy because of the FDA's retaliation against Barr.

And beyond calculation was the damage done to Marvin Seife, whose only crime was camaraderie. As someone who knew him said later, "Whatever his sentence was, it wasn't amputation."[33]

5

Toys 'R Us

IN 1976, CONGRESS PASSED THE MEDICAL DEVICES AMENDMENT TO THE basic Food, Drug, and Cosmetics Act, designed to ensure the safety and effectiveness of medical products as diverse as heating pads, joint implants, and intrauterine devices, or IUDs. The amendment required manufacturers to register with the FDA, to follow quality control procedures, and in some cases to gain premarketing approval from the agency. Roughly speaking, the passage of the amendment brought the approval process for devices into line with the process for pharmaceuticals, and was prompted, in part, by the disastrous effects of one particular IUD, the Dalkon Shield. Just as it took the sulfanilamide killings of 1937 to prompt the passage of the 1938 FDC Act, just as it took the furor over thalidomide to prompt the passage of the Kefauver-Harris Amendments, it took the tragedy of the Dalkon Shield to secure the devices amendment.

The first use of an intrauterine device as a means of contraception is lost in antiquity. For centuries, before setting out on a long journey across the desert, the camel drivers of the Middle East inserted pebbles into the uterus of a female camel to keep her from becoming pregnant during the trip. A herdsman's trick, tested by time, and it worked.[1] Although no one knows exactly why, a foreign body placed in the uterus will prevent pregnancy most of the time. According to Richard B. Sobol, "One theory is that in reaction to the device the body produces white cells that destroy the sperm or the fertilized egg. Another is that the IUD somehow interferes with the ability of the fertilized egg to attach itself or to remain attached to the wall of the uterus."[2]

Historically, devices made of ivory, wood, silver, gold, copper, and zinc have been used for this purpose. But whatever the substance, injury and infection have always been associated with IUDs, and they were not widely used until the late 1960s when they were recognized by many women as a valid alternative to the Pill. The IUD was almost as effective, and it did not introduce synthetic hormones into the body. And once in place, women could not forget to use it, which made it seem particularly appropriate for young or poor women, two groups traditionally viewed as being uncooperative and irresponsible by population control advocates. At the height of their popularity, IUDS were manufactured in more than a hundred shapes and sizes, and were used by more than 60 million women worldwide.[3]

One such device was the Dalkon Shield IUD, designed in 1968 by Hugh Davis, a gynecologist on the faculty of the Johns Hopkins Medical School, and Irwin Lerner, an electrical engineer. One design problem that Davis and Lerner had to overcome stemmed from the natural tendency of the body to rid itself of a foreign object. Because of that tendency, IUDs were often expelled involuntarily, and the Dalkon Shield was designed with that problem in mind. Made of a piece of flexible plastic, the device had either four or five prongs jutting out and downward from each side. The function of those prongs was to resist expulsion, but in practice they often embedded themselves in the uterine wall with unpleasant, and sometimes disastrous, results.

In addition, Davis and Lerner made a fundamental mistake in the material that they chose to use for the tail strings of the device. All IUDs have a tail string attached to the bottom of the device, passing through the cervix into the vagina. Its purpose is to allow the woman to check that the device is in place, and to aid in its removal. Every IUD other than the Dalkon Shield used a tail string made of a single plastic filament in order to prevent the absorption of moisture and bacteria from the vagina into the uterus, a major cause of uterine infections.

Instead, Davis and Lerner used a multifilament string composed of hundreds of fine nylon strands enclosed in a nylon sheath. They hoped that the sheath would prevent vaginal fluid from seeping up between the strands and into the uterus, a process called "wicking," but, inexplicably, they failed to seal the sheath at both ends. In fact, the sheath accentuated the problem by shielding the bacteria wicking upward from the antibacterial fluid that occurs naturally in the cervical plug. The result was a device that provided a safe conduit for bacteria into a uterus already torn, or even perforated, by the action of the prongs, thus creating a fertile breeding ground for pelvic inflammatory disease, or PID. The consequence of PID is often a hysterectomy, and even when that procedure is not required, infertility often results.[4]

In October 1969, Davis submitted a study of the Dalkon Shield to the *American Journal of Obstetrics and Gynecology*. Although seriously flawed by statistical inaccuracies and outright deceptions, the study was published in January 1970, and shortly thereafter the rights to manufacture the device were sold by Davis and Lerner to the A. H. Robins Company, a major pharmaceutical house, for $750,000 plus 10 percent of the net sales, a substantial sum for the times.[5] Had this been after 1976, the Dalkon Shield would then have been subjected to the same rigorous testing, in theory, that a new drug would have undergone before being allowed on the market, a scientifically sound clinical testing of 1,500 users over a two-year period. *In theory* because, as we shall see, even after the passage of the Devices Amendment, the regulation of the devices market by the FDA was considerably less than rigorous. In any event, the year was 1970, and the Robins Company was able to manufacture and market the Dalkon Shield with no more government regulation than if it had been selling pots and pans.

Problems with the Dalkon Shield became apparent at once. The prongs tore at the uterus, the tail strings channeled bacteria, and, as a final insult to the consumner, the device didn't work all that well. As Sobol reports:

> During the years that Robins marketed the Dalkon Shield, the company was constantly engaged in stemming the tide of adverse information—information reporting high pregnancy rates and information indicating the hazards of the product. Ironically, the two problems intersected, because the first reports of serious injuries associated with the Dalkon Shield concerned women who had become pregnant with the device in place and had suffered septic abortions. A septic abortion is a spontaneous termination of a pregnancy resulting from an infection of the reproductive system.[6]

During 1972 and 1973, Robins received a flood of reports involving spontaneous septic abortions, along with reports of other uterine infections among nonpregnant women using the shield. Although some of the women had come close to death because of these infections, Robins did nothing to address the problem, not even when two young women who had been wearing the shield became pregnant, underwent spontaneous septic abortions, and died. It was only in October 1973 that Robins equipped the device with a new label conceding the possibility of "severe sepsis with fatal outcome, most often associated with spontaneous abortion following pregnancy with the Dalkon Shield in situ."[7]

More deaths followed, more complaints, and more unwanted pregnancies, until the wave of negative publicity forced Robins to withdraw the Dalkon Shield from the domestic market. In July 1974, the Centers for Disease Control in Atlanta specifically named the Dalkon Shield as being associated with septic abortions, and in April of the following year Robins finally took the product off the market worldwide. By that time 15 fatal and 245 nonfatal septic abortions had been reported, some 3.6 million Dalkon Shields had been implanted in women all over the world, and most of those were probably still in place. Even then, Robins, displaying the bravado of a corporation bracing to receive a spate of lawsuits, insisted that it "remains firm in the belief that the Dalkon Shield, when properly used, is a safe and effective IUD."[8]

The lawsuits, when they came, were defended by Robins and its lawyers with all the finesse of a back-alley brawl. The first case was tried in December 1974, and the battle lines were drawn when the jury returned a verdict for the plaintiff, Connie Deemer, in the amount of $10,000 in compensatory damages and $75,000 in punitive damages. Deemer had become pregnant with a Dalkon Shield in place. The IUD perforated her uterus and lodged itself in her abdominal cavity. Although she successfully caried her baby to term, she later was forced to undergo surgery to have the device removed.

Robins had defended the Deemer case in a fairly routine fashion, but once the verdict was in it was no more mister nice-guy. Robins' insurer, the Aetna Casualty & Surety Company, assumed the direction of all Dalkon Shield litigation, and engaged the Richmond law firm of McGuire, Woods & Battle to restructure the basic defense.

The new game plan argued that there was nothing wrong with the Dalkon Shield, and that no particular plaintiff could prove that her injury was caused by the shield, and not by some other factor. A panel of experts was assembled to testify that the device was in no way defective, or materially different from other IUDs. Thus, went the reasoning, any infection suffered by the plaintiff must have been due to some other source.

To support this defense, says Sobel, "Attorneys for Robins would interrogate plaintiffs about their sexual and hygienic habits. Robins took the position that multiple sex partners, with the accompanying increased risk of contracting sexually transmitted diseases, were a more likely cause of uterine infections than was the Dalkon Shield."[9]

At that point the lawyers rolled in the heavy artillery. The plaintiff was asked to identify all of her sex partners, on the slippery ground that these men should be subpoenaed and asked about their medical histories. The women were also asked to describe their sexual practices and the details of their personal hygiene, with the implication that one or the other of these could have caused a uterine infection. Some of the courts involved did not allow these answers in evidence, but others did, and the simple fact that the questions could be asked dissuaded many women from bringing any action at all, and prompted others to accept small settlements from Robins.

The role played by the FDA in the Dalkon Shield episode was minimal at best. The position of the agency, prior to 1976, was that its charter did not call for the regulation of medical devices, and thus it had no more authority in the matter than, say, the Department of Agriculture might. But this excuse was both self-serving and inaccurate. The FDA had the power to empanel experts to hear testimony on all medical matters, including devices, and in June 1974 such hearings were actually held on the subject of septic abortions and IUDs.

At those hearings the medical director for Robins argued that the risks presented by the Dalkon Shield were no different than those of other IUDs, but he carefully did not reveal that the Dalkon Shield used a multifilament, not a monofilament, tail string disguised within a nylon sheath. This was information unknown to the FDA and to the medical community at large, although a simple examination involving a tool no more sophisticated than a pocket knife would have revealed it. No one, apparently, had a pocket knife, and the sole agency action was a request to Robins to suspend sales while the safety issue was studied further. Two days later, Robins was able to announce a "voluntary" suspension of sales, while stating that, "Neither A. H. Robins nor the FDA has any reason at this time to believe that women now using the Dalkon Shield successfully should have the device removed."[10]

Further FDA involvement was equally inadequate. In 1979, a Colorado jury awarded $550,000 in compensatory damages and $6.2 million in punitive damages to a woman who had become pregnant while wearing a Dalkon Shield and had suffered a near-fatal septic abortion. It was, at the time, the largest judgment ever imposed on a pharmaceutical company. In 1980, another jury awarded $600,000 to a woman who had undergone an extrauterine pregnancy while wearing the shield, and who later had all of her reproductive organs removed. That same month Robins paid $1.3 million to settle a case involving a child born with brain damage linked to an infection caused by the use of the shield. All told, in 1980 almost a thousand new Dalkon injury cases were filed, many of them based on injuries that had occurred after Robins had stopped selling the product.

Clearly, there was not a woman in the world who should have been wearing the device at this point, and the lead attorney for a group of the plaintiffs, Bradley Post, appeared before an FDA panel to ask that the agency recommend the removal of all Dalkon Shields. For reasons known only to its members, the panel chose to remain unconvinced by the mass of scientific evidence that supported a general removal, and declined to take any action at all. Three years later, Post and another attorney, Robert Manchester, petitioned the FDA to initiate a recall of the shield and to require Robins to pay for the costs of removing the devices. Once again, the FDA did nothing.

If a defense of the agency's inactivity and indifference to the fate of more than three million women is to be mounted, it can only be said that prior to the passage of the Devices Amendment in 1976 the FDA lacked a clear regulatory mandate in the case of the Dalkon Shield. But for a regulatory agency to perform its functions honestly and impartially, attention must be paid to more than the letter of the law. In the case of the Dalkon Shield, the FDA had a collective moral obligation to comment openly, to urge, and if necessary to coerce Robins into revealing the flaws that made its product so dangerous to so many women. Instead, the FDA played the industry's game of delay, obfuscation, and a backing away from responsibility.

In the end, the lawyers for Robins were able to maneuver federal bankruptcy law to halt six thousand lawsuits in courts throughout the United States, and to secure, within the friendly confines of a hometown court, a pair of decisions limiting the pool of women eligible for compensation, and at the same time limiting the company's liability to an estimated $2.4 billion. The court approved a reorganization plan that permitted Robins' shareholders to receive four times the prebankruptcy value of their stock, free of the claims of the injured women, and at this late date there are still grave doubts that the compensation fund will prove sufficient to pay the claims of the approximately two hundred thousand women recognized by the plan.[11]

Clearly, the responsibility for the Dalkon Shield tragedies could not be laid exclusively at the doorstep of the FDA. There were villains enough in the story: Davis and Lerner, A. H. Robins, their lawyers, their insurance carrier, and the hometown judge who effected their cozy settlement. The FDA was, at worst, a passive policeman with an empty pistol, lacking the authority to regulate effectively. Give us the power, was the agency line, and let us do the job. The Devices Amendment of 1976 endowed the FDA with just that power and responsibility, but, regrettably, the passive policeman remained on the beat. In the case of the Bjork-Shiley heart valve, hundreds of lives and millions of dollars were lost while the policeman looked the other way.

Imagine 55,000 people engaged in a massive version of Russian roulette, all of them walking around with the medical equivalent of time bombs embedded in their hearts. Not all of those bombs will go off, not even most of them, but some of them will, and in the best tradition of Russian roulette, not one of them knows whose turn will be next.

Tony Luizzo, of Brooklyn, New York, was one of the 55,000. Tony was twelve years old when an artificial valve was implanted in his heart, a Bjork-Shiley 60-degree valve manufactured by Shiley Inc. of Irvine, California. The implantation was successful, and over the years Tony learned to live with the faint ticking sound that the valve gave off as it opened and closed, controlling the flow of blood in his heart. As an adult, Tony remembered that, "It never bothered me. It was sort of comforting. As long as I could hear the ticking, I knew that my heart was working."[12]

That comforting feeling disappeared in February 1990, while Theresa Luizzo, Tony's mother, was watching a television program about artificial heart valves, and one type of valve in particular. Since 1982, said the narrator of the program, there had been disquieting reports of deaths involving this valve, reports that were out of proportion when compared with other valves. These particular valves were breaking up inside the heart, and people were dying. The valve was the Bjork-Shiley 60-degree.[13]

The valve received its name from its shape and the size of its opening angle. Consisting of a convexo-concave disc inside a metal ring that was covered with Teflon, the valve was sutured to the heart to hold it firmly in place. The valve opened to a 60-degree angle, and was supported by two wire holders, the inflow and the outflow struts, located on each side of the disc. The inflow strut was an integral part of the ring, while the outflow strut was a separate piece of wire that was welded on. In every case of valve failure reported, the outflow strut had been the culprit. When the outflow strut fractured, the disc escaped from the ring, thus causing an uncontrolled flow of blood through the heart. In two-thirds of the cases reported, death followed quickly.

Theresa Luizzo recalls how her life turned upside down that night. "To hear something like that on a television program was incredible. We'd never heard a word from Shiley, or from the doctors, and this had been going on for years. It came like a bombshell."[14]

That bombshell, like Tony's valve, had been ticking for some time. When Congress passed the Medical Devices Amendment in 1976, it gave the FDA, among other things, the authority to inspect manufacturing operations and to track product performance. It also charged the agency with the writing of certain regulations that applied to the manufacturers of devices considered to be "critical." The very word *critical* indicated a compelling need for timeliness, but given the history of the FDA in the writing of regulations, it was no surprise that ten full years went by before these vital rules governing the Premarket Approval Process for critical devices were written.

Dr. Sidney Wolfe, head of the Public Citizen Health Research group in Washington, D.C., an outspoken critic of the FDA, has pointed out that, "It took the FDA all those years to draw up the regulations saying that if someone is killed by your product, you have to report it. So, technically, from 1976 to 1986, if a company didn't want to report deaths or injuries due to a device, they didn't have to."[15]

And there were deaths. The troubled history of the Bjork-Shiley 60-degree valve had begun before the device ever reached the market when the first of many fractures occurred during clinical trials. Despite this failure, the FDA approved the device in April 1979, shortly after Shiley was acquired by the giant pharmaceutical house of Pfizer, Inc., and the valve was admitted into the open market. It never should have been, and problems surfaced quickly. Between 1980 and 1983, the valve was recalled by the manufacturer three times because of fractures, and three times a supposedly improved version was returned to the marketplace, but the fractures continued. During much of this time the FDA played the familiar role of the passive policeman, although there were those within the agency who were obviously disturbed by what was going on. The situation created a rift within the Center for Devices and Radiological Health (CDRH). On one side were the proponents of scientific advancement, who saw the Shiley valve as a technologically elegant piece of design with regrettable tendencies that would be taken care of in time, and on the other were the members of the enforcement field staff, who watched with dismay as the number of deaths mounted.

The dynamic between techonology and enforcement can be, and should be, healthy, but in this case it paralyzed the agency. Enforcement floundered, and hard decisions were avoided. Twice, in 1982 and 1984, the FDA asked Shiley to halt the production of the valve on a voluntary basis, and when Shiley refused, the agency did nothing about it. Although the FDA did not, and still does not, have the power to order a recall, the agency could have withdrawn its approval for marketing the valve, which would have accomplished substantially the same purpose. The agency did not, and years later in an appearance before Congress, James Benson, then the deputy commissioner of the FDA, could explain the lack of action only by saying, "It was a case of taking things in order in the queue."[16]

But there were others who thought that the queue should have been jumped. "By the middle of 1984," said Dr. Wolfe, "it was clear that this valve was uniquely dangerous. We petitioned the FDA to have it taken off the market, and the very next day I got a call from a guy named George Sherry."[17]

Sherry was a former Shiley employee who had been in charge of the drilling and welding operation on the valve from January 1982 until September 1983, when he left the firm in protest over manufacturing procedures. He was ready to blow the whistle, and in a series of meetings and memos he laid bare to Wolfe the severe problems involved in the manufacture of the valve. Wolfe recalls that, "He supplied us with internal documents from the company showing that everyone there had known for years exactly what was wrong, and the engineering drawings that showed that the valve was fatally designed."[18]

According to Sherry, the product drawings had major errors and conflicts, and the valve could not be built if the drawings were to be followed. The drawings were changed haphazardly, some with as many as fifteen changes on them. "It was apparent to me that the method used was like sitting back and throwing darts at a wall."[19]

Shiley did not use certified welders on the job, and the training of their welders was questionable. Some struts had to be rewelded three and four times in order to get a proper result. In addition, the mass of the welding fixture was too great, thus requiring amperage levels of 40 and 50 amps instead of the recommended 12. As a result, the temperature of the weld joint rose much too quickly, and thus impaired the strength of the joint.[20]

Holes installed in the flanges of the struts were often drilled in the wrong places, but instead of discarding such pieces, the struts were bent so that the discs could be forcibly inserted.[21]

Given such manufacturing procedures, it was not surprising that the outflow valve was breaking up inside the patient with a deadly regularity, but Sherry's superiors seemed unconcerned. In a series of meetings with his supervisors, Sherry detailed a list of corrective measures needed to be taken in order to avoid strut fractures, but few of these changes had been implemented by the time he left the company. He recalls that, "I had virtually no success in bringing the welding procedures up to an acceptable level. I was totally frustrated."[22]

Adding to his frustration was the attitude of the FDA during this time. Given the fact that the agency conducted a series of inspections at Shiley during the period when the struts were fracturing, and given the fact that all of the fractures had occurred at the welded joint, why was it that not one FDA inspector ever asked Sherry, the man in charge of the welding operation, what he thought might be going wrong? It was at this point that the Center for Devices began to be known in certain Washington circles as "Toys 'R Us."

On July 11, 1984, and August 8, 1984, Wolfe's organization, Public Citizen, armed with the information received from Sherry, presented the FDA with information on strut fractures of the Bjork-Shiley valve that clearly warranted removing the device from the market. In a letter to FDA Commissioner Dr. Frank Young, Wolfe pointed out that, "Ninety-six strut fractures have been reported to FDA, with two-thirds of them resulting in patient deaths. The actual number of strut fractures and deaths may be five times higher."[23] Wolfe went on to point out that despite the known shortcomings in manufacturing procedures, and despite the knowledge received from Sherry of the specific steps needed to remedy the defects, the FDA had failed to require Shiley to improve the manufacture of the valve.

"We went public with it," Wolfe recalls, "and we thought at that point that the FDA would take the device off the market, but they didn't. They stalled around for another two years, and all that time people were dying."[24]

While they were dying, neither Shiley nor the FDA did anything constructive about notifying people like Tony Luizzo that a potentially defective device had been implanted in their hearts. The manufacturer did send a series of "Dear Doctor" letters that advised implanting surgeons of the estimated risk associated with its valves, but nothing was sent to the patients or to their family doctors. Most patients remained in total ignorance of the situation for years, and even those who were aware of the facts could do little to help themselves. For most recipients, Tony Luizzo among them, the option of replacing the valve with one of a different make and model did not exist, for their hearts could not be subjected to any further surgical stress. Added to that was the financial burden that each recipient would have to assume, with insurance companies holding that the replacement of a valve that had not yet failed would be considered "cosmetic" or "elective," and would not be covered. As one recipient put it with dark humor, "The best I can do is find a house next door to a hospital with a first-class emergency room and a cardiac surgeon on duty at all times."[25]

But there was little humor to be found in the situation. Called before Dingell's Subcommittee on Oversight and Investigations, FDA officials were pressed to explain why the agency had moved so quickly to approve the Shiley valve in the absence of good clinical data, yet had moved so slowly in instituting regulatory action. There was no good, true, or honest answer to those questions. Caught between science and common sense, the agency had failed to heed the warnings of its own engineers and compliance officers. The FDA had the effective power to suspend Shiley's operations until the problem with the valve was identified and addressed. Instead, the agency allowed the product to be produced, sold, and implanted until 1986 when Pfizer and Shiley, faced with an inundation of lawsuits, finally withdrew the valve from the market.

But the fractures continued. Hundreds of valves went bad, and hundreds of people died, but even those numbers were assumed to be understated since the distress symptoms of patients with common heart seizures and those with fracture problems were almost the same, and autopsies were not always taken. And as the number of deaths mounted—on the average of 7 out of every 10,000 recipients each year[26]—the number of lawsuits mounted as well. Out-of-court settlements with Pfizer were made by the families of some of the victims, and a variety of class-action suits were brought by those whose valves had not yet failed. Through all of this, the official line at Pfizer, as expressed by its chief executive, William C. Steere, was that, "These people were going to die anyway, and they are alive today because of the valve."[27]

But every stonewall must come to an end, and in December 1991, Pfizer sold off Shiley Inc. to a subsidiary of the Italian FIAT Group, while retaining its liability for the defective valves. Then, in an effort to put the whole messy business behind it, the pharmaceutical giant proposed the creation of a fund of $80 million to $130 million to pay patients with the valve for cardiac consultations, and offered to set aside an additional $75 million for research to identify those recipients who still did not know that they were at risk.[28]

The offer was amended as part of the settlement of a class-action suit in which Pfizer set aside $75 million to pay for research and development of methods to detect cracks in the valves before the valves break. The money was also earmarked to pay for open-heart surgery to replace valves already cracked and, in addition, Pfizer established a separate fund of $300 million to compensate patients and their families.[29]

But for all of that, the world's biggest game of Russian roulette went on, and Tony Luizzo no longer found any comfort in the ticking of his valve.

"It's different now," he said. "I keep waiting for it to stop."[30]

In the chaotic days just after World War II, the U.S. Navy in occupied Japan was struck by a mysterious rash of dockside

burglaries. The mystery lay not in the burglaries themselves—
dockside thievery is as much a part of the maritime life as the
winds and the tides—but in what was being stolen. The Shore Patrol
reports indicated that along with such highly prized items as
canned hams and whisky, the thieves were concentrating on an
innocuous commodity of little apparent value. For some unknown
reason, the hottest seller on the black market was now the liquid
insulation for ordinary electrical transformers, an item that the
Americans had shipped by the thousands to Japan in an effort to
repair that nation's bombed-out electrical network. Why this liquid,
the authorities wondered? Who would want it—what could be done
with it? It was, after all, nothing more than simple silicone.

The answer was as simple as the substance itself. The demand
for silicone was coming from the Japanese prostitutes who worked
the port areas, and who had quickly realized that American
servicemen preferred women with larger breasts than was
common among Japanese women. Hurried consultations with
waterfront doctors had led to silicone injections, and would lead in
turn to a medical problem that would plague the FDA forty years
later.

Silicones are chains of silicon molecules with side groups of
other molecules. Silicones are versatile, as the molecules can be
linked up into short chains, which make a runny liquid, or into
longer chains to make solid materials such as rubber bands and
blocks. (In later lives, the short-chain version would emerge in the
1950s as "Silly Putty," and in the 1990s as "Gak.") For all of its
versatility, however, silicones were little more than laboratory
curiosities until World War II when the government commissioned
Corning Glass and Dow Chemical jointly to find a substitute for
badly needed rubber. Synthetic rubber was eventually developed,
but one of the first forms of silicone to come out of this effort was
an insulator liquid for electrical transformers, the liquid so in
demand in postwar Japan.[31]

The doctors who catered to the Japanese prostitutes first
experimented with injecting a variety of substances into the

women's breasts in order to increase their size. Goats' milk and paraffin were early favorites, but silicone soon became the product of choice, and it was then that the transformer fluids began disappearing from the Navy storehouses. It was also then that the problems began. American plastic surgeons picked up the Japanese practice of injecting liquid silicone directly into the breast, and reports soon began to reach the medical establishment of women who had received the treatment, women with lumpy and ulcerated breasts, and problems in the chest, back, and arms that came from drifting bits of silicone.

What was clearly needed was an "envelope" to contain the silicone and keep it from wandering through the body, and in 1962, Dr. Frank Gerow and Dr. Thomas Cronin, who later joined Dow Corning, combined the properties of the rubbery form of silicone and the liquid form to make a gel that was soft but firm. This they wrapped in an envelope of the rubbery form, the elastomer, to produce the first true silicone gel breast implant. Other forms of implants would be developed over the years employing materials such as saline and foam, but silicone gel was the basic device for more than three decades.[32]

The silicone gel breast implant differs from the other two devices discussed in this chapter in that it is difficult to make a sharply defined, black and white case against the use of the product. There is no doubt that the Dalkon Shield killed and maimed many women. There is no doubt that the Bjork-Shiley 60-degree heart valve contributed to the deaths of hundreds and hundreds. And there is no doubt that many women have suffered a wide range of adverse reactions after receiving a silicone implant.

According to studies at the University of Texas dating back to the 1970s, the human immune system reacts to the silicone in breast implants by making antibodies against it. The antibodies then attack the silicone and whatever tissues are associated with it, with arthritis, scleroderma, and lupus as some of the consequences.[33] Joint pain, rashes, and flu symptoms also result. But the fact remains that millions of women around the world have had these implants and are delighted with them. (Fully 20 percent of the implants go not for cosmetic effect but for reconstructive purposes to women who have had cancer surgery.) Representatives of these women, often subsidized by the implant makers, have been the loudest in the defense of the device.

But those points having been made, any discussion of breast implants tends to become involved in cultural, rather than medical, matters. Women, understandably, tend to be more vocal on the subject than men, but there is no unified female front. For some women, the essential point is the right to do with their bodies as they wish without paternalistic intervention or regulation. Lacking evidence that silicone implants are dangerous to them specifically, they argue for the right to make an informed choice. On the other side are the anti-implant women who argue that their bodies are being recklessly endangered by a male-dominated medical establishment that is indifferent to their concerns.

Both sides, however, agree with something close to regret that the American culture, almost uniquely, is obsessed with the concept of the large and nurturing bosom. Part of this, they say, is obvious momism, but part is also a perceived sexuality. Large breasts in America are not only a sign of pulchritude but they are also seen by a significant sector of the male population as a sign of sexiness. Not sexiness in the eye of the beholder but in the nature of she who is being beheld. Large-breasted women, goes the theory, enjoy sex more than their small-breasted sisters. The knowledge that these concepts arise among semiliterate males of the great unwashed area does little to ease the concerns of these women, mainly because they know that there are just as many other women across the country who have been conquered by the concept, women who have accepted a locker-room joke as an article of faith.

Examining the point, an editorial in the *New York Times* quoted a woman who said of her own cosmetic surgery, "You know, it means so much for a woman not to be small, not to feel disfigured because God didn't make her enormous." To which the *Times* replied rhetorically, "Who made you think you were disfigured? And why did you believe them?"[34]

The scientific community, as a whole, also has had a hard time coming down on a single side of the subject. For every qualified opponent of implants there is invariably another (also often subsidized by industry) willing to minimize the risk involved with the rationale that, "Just because penicillin sometimes causes anaphylactic shock doesn't mean that you take penicillin off the market." Those who do not believe that silicone can cause true autoimmune disease admit there is no doubt that foreign material in the immune system can cause some reaction, but they say this reaction is mostly local and mostly a response of cells like macrophages, whose job is to devour foreign materials.[35]

But even the optimists admit that there are questionable hazards involved. The insertion of the implant often causes a strong reaction that prompts the body to construct a container of fibrous tissue around the device, walling it off from the rest of the system. This condition, capsular contracture, can be painful and unsightly, and can make the breasts hard. Researchers say that the condition occurs in varying degrees in about 10 percent of the women receiving implants. A more deadly hazard is that silicone is opaque to X-rays, thus making mammograms to find early breast cancer very difficult. Studies show that on average, by the time that cancer is discovered in a woman with implants, the tumors are five to six times the volume of tumors found in women without implants. Further studies show that 45 percent of women with implants had cancer that had already spread to the lymph nodes at the time of diagnosis. The figure for women without implants whose cancer had been discovered by mammography was only 6 percent.[36]

All of this was in the future when the first silicone gel implants came on the market in 1963, but warning signs soon appeared, signs that should have at least hinted at the hazards involved. Breast implants were not the only medical products developed from silicone. Surgical clips, penile prostheses, intraocular lenses, and tubing for blood oxygenators and dialysis machines were some of the others, and although these devices worked well, they sometimes caused problems. Patients on kidney dialysis machines using silicone tubing were found to develop liver disease at an unusually high rate, and subsequent autopsies showed a large number of silicone particles in the liver. The problem disappeared when different tubes were used.

Similar problems emerged during cardiac bypass surgery when silicone was used as an antifoam agent in the oxygenation of the patient's blood. Another warning was the incidence of scleroderma, an autoimmune disease that, according to critics of implants, is linked to the devices. Scleroderma, in which the growth of fibrous tissue leads to a thickening of the skin, has historically been associated with silica or silicon, the natural element from which silicone is derived. Of 120 men with scleroderma studied in Germany between 1981 and 1988, 93 of them had direct exposure to silica as miners, sandblasters, or in similar occupations.[37]

So the warning signs were there, but aside from making and selling implants as quickly and as profitably as possible, what was industry doing about them? Actually, very little. Studies of the device were kept to a minimum, and when they were actually implemented the results were often either concealed from the public or misrepresented. In a 1973 study involving thirty-eight dogs, Dow Corning scientists found that virtually all the animals showed inflammatory reactions to silicone. Four dogs were given gel implants, and of those one showed damage to internal organs and died, another formed a benign tumor, and two developed inflammations that lasted more than two years. The official Dow Corning report, however, showed none of this. According to the report, the health of all the dogs remained normal, and only slight inflammation was seen.[38]

In a similar case of concealment, Dow kept buried for twenty years a report from its own researchers that voiced concern over leaking implants that could send silicone floating through the body with undetermined consequences.[39]

Dow's manufacturing procedures were also questionable to the point of criminality. For years the company concealed a quality-control problem that occurred at a manufacturing stage just after the implant's bag had been filled with silicone and sealed. The implant was then placed in an oven for a process called polymerization that would turn the contents into a substance that more closely approximated the consistency of natural breast tissue, and would make it less likely to leak from the bag. The quality control was a paper strip from an automatic recorder that registered the temperature levels in the oven, but whenever the ovens did not work properly the Dow technicians would destroy the paper strip and substitute a false record showing that all had gone well.[40]

Years later, when the subterfuge was exposed, a high Dow Corning executive was asked for his description of the practice. He answered, " 'Faked it' is a fair description of what they did."[41]

The concern of rank-and-file Dow Corning employees over the product they were making and selling was made clear in the internal memoranda that eventually surfaced decades after they were written.

In a 1976 note, engineer Tom Talcott asked, rather plaintively, "When will we learn at Dow Corning that making a product 'just good enough' almost always leads to products that are 'not quite good enough?' "[42]

In 1978, a Dow salesman in the Detroit area reported that four plastic surgeons were experiencing ruptures of the implant about 15 percent of the time, and wondered, "Are we making the envelope different, and is it weaker?"[43]

In 1977, a Dow marketing executive complained that the company was only halfheartedly looking into allegations that silicone "bleed" was contributing to the formation of a fibrous tissue around the implant. "I know," he wrote, "of at least one loyal Dow Corning customer who believes that our prosthesis bleeds more than other gel prostheses, and is considering shifting to a competitive product."[44]

The competition referred to came from four other manufacturers of implants. Two of these, Mentor Corporation and McGhan Medical Corporation, produced not only silicone gel implants but implants that used a saline solution within the envelope. The advantage of saline as a filler is that if the implant bursts accidentally the fluid is absorbed into the body without any adverse effects. The disadvantage is that saline implants can deflate spontaneously, and often need to be replaced. Thus, silicone continued to be the implant of choice.

Writing in the *Wall Street Journal,* Mayo Clinic Doctors John E. Woods and Phillip G. Arnold explained their preference for silicone. "While saline implants can be used for both augmentation and breast reconstruction, gel-type implants are superior, resulting in softer breasts with more normal contours. It is inappropriate that thousands of women should be denied the opportunity for optimal results because of a very small minority of women with problems."[45]

The third of Dow Corning's competitors was Bioplasty, which also made a silicone gel implant, and the fourth was the Markham implant, which was marketed under such brand names as Meme, Natural Y, and Replicon. The Markham product became an industry leader in the 1980s, and was acquired in 1988 by Bristol-Myers Squibb. This implant was also a bag of silicone gel, but with an added feature, an outer layer of spongy polyurethane foam designed to prevent the hardening of breast tissue. In company documents this foam was described as the "patented Microthane interface system," but it also had another name: Scott Industrial Foam. This was the foam used in automobile air filters and carpet-cleaning equipment, and in the human body it can degrade into a substance called 2,4-toluene diamine, or TDA, a chemical that is suspected of causing cancer in humans.

TDA is classified by the government as a hazardous waste, and was banned for use in hair dyes in 1971. Workers who handle it are advised to wear goggles, rubber gloves, and respirators, and it has been found in the breast milk and the urine of women with polyurethane-coated implants. The foam's manufacturer, Scotfoam Corporation of Eddystone, Pennsylvania, never intended their product for medical purposes, and in 1987 they informed the implant manufacturer that the foam should not be used in people. The reaction of the manufacturer could have been anticipated. With little or no government control over breast implants, the foam continued in use all through the 1980s, effectively turning the bodies of many women into toxic waste dump sites.[46]

To anyone familiar with the history of the FDA in the 1980s, its lack of action in this area, its indifference to the problem, indeed its reluctance to admit that a problem existed, comes as no surprise. The agency defense is that when breast implants first came on the market in 1963 the FDA had no regulatory authority over medical devices. But that defense falls apart after the passage of the Device Amendment in 1976. A clause in that amendment called for existing devices to be classified as either needing or not needing a retroactive safety check. At that point a dispute developed within the agency over whether breast implants should be classified as a "Class 3" device, which would have required a safety study, or a "Class 2," which would not.

The dispute continued unresolved for twelve years, not an FDA record for procrastination, but high up on the list. From 1976 through 1988, while thousands of women were suffering the adverse effects of silicone implantations, the FDA pondered the need for a safety study. Committees met, memoranda were exchanged, reports were written, and absolutely nothing was done to resolve the question. Finally, in 1988, with lawsuits beginning to mount against the manufacturers, the agency began the lengthy regulatory process of asking for safety data, and it was not until 1990 that the FDA actually requested that the manufacturers file for formal approval of their products.

Device evaluation is a much younger science than drug evaluation, and when the FDA was first confronted with implants, no one either inside or outside the agency knew enough about silicone to know what questions had to be asked. All that was known at the time was that silicone had been used apparently successfully in a variety of medical devices, from pacemakers to the lubricants on syringes. The substance was stable, resistant to body fluids, and was not rejected by body tissues. But certain basic questions had to be asked, and the FDA never asked them.

The agency never asked Dow, or any other manufacturer, to test silicone over a long term in humans.

The agency never asked for a study of silicone's immunological effect on the body.

The agency never asked for the establishment of a registry of women who had received the implants, a procedure that would have enormously simplified the basic research on the subject.

The agency did next to nothing until public opinion and mounting lawsuits forced it into action. And even then, after twelve years of negligence, it took another four years to hold its first public hearings on the subject.

The problems that plagued the CDRH continued through the end of the decade, well into the 1990s, and still have not been resolved. Even after a younger, more dynamic commissioner had taken charge of the agency, the Center continued to come under fire from critics both in Congress and in industry for its extensive backlog and delays in bringing promising new products into the marketplace. One congressional report cited a backlog of more than 1,100 applications for approval of new devices, not including a sizable number of applications sitting unopened in the mail room.[47]

The beginnings of reform finally came after the establishment and a report of the old reliable blue-ribbon panel, and the appointment of a new director for CDRH. But in the 1980s that was still part of the future, and it was business as usual for Toys 'R Us.

6

Vitamins, Hurricanes, and
Killer Grapes

MOST OF THE FDA STORIES THAT MADE THE HEADLINES DURING THE 1980s dealt with matters of moment: lifesaving drugs, tainted blood, defective devices. But there were other, quieter matters of equal importance that showed the agency at its best, and at its worst.

During the decade there was a running three-way struggle over vitamins and minerals with no holds barred. On one side were the vitamin libertarians, those who insisted on the right to take doses of nutrition supplements of any size and any combination without interference from government agencies. On the second side stood a body of nutritionists who urged the public to have nothing to do with manufactured supplements, and to rely instead on a well-balanced diet. And on the third side were the regulators of the FDA, whose ostensible purpose was to prevent the manufacture and sale of unsafe products, but who were ideologically aligned with the nutritionists to the point that, had it been legally possible, the agency would have banned the sale of supplements altogether.

It is kindest to assume that this institutional prejudice at the FDA against the use of dietary supplements evolved from ignorance rather than from some inherent aversion. Since medical schools until quite recently did not teach nutrition, the very lack of knowledge about nutrition was taken as an adequate basis to preclude dietary supplements from the diet. Since the mainstream medical literature discussed neither nutrition in general nor the use of dietary supplements for the prevention of long-term disease, a strong policy against the use of dietary supplements grew and flourished at the FDA.

The lines of the vitamin combat were sharply drawn, and the battle was fueled by confusion and misinformation. When a front-page story in the *New York Times*[1] inaccurately reported that "armed agents" of the FDA had raided an alternative medicine clinic in Kent, Washington, the vitamin libertarians rose up in righteous indignation, determined to defend what they considered to be a constitutional right. A retraction by the *Times*[2] one week later did little to ease the situation. The serious vitamin users were convinced, with some justification, that the FDA was bent on harassing the supplement industry, the FDA insisted that it was only trying to enforce the law, and a tertium quid of nutritional scientists tartly observed that the use of chemical vitamins produced nothing more than expensive urine.[3]

As usual, the FDA was twenty years behind the times, vainly trying to stem a tide of public opinion that enthusiastically endorsed the use of dietary supplements. By the end of the 1980s, vitamins had come to play a major role in the everyday health care of 80 million Americans;[4] they were no longer the innocuous tablets taken every morning with the vague understanding that they were "good for you." At one time consigned to the status of supporting actors by mainstream medicine, vitamins and minerals had moved to stage center as more and more research showed that the influence of these nutrients on the body was far more extensive than had been suspected. And that influence had been translated into dollars. By the end of the decade, vitamins and minerals were going over the retail sales counters at the rate of more than $3 billion each year,[5] and what once had been a cottage industry patronized mostly by advocates of alternative medicines had grown into an industrial fortress well worth defending.

What they were defending was the right to take into their bodies those carbon-containing substances that are required for normal metabolism, but that are not synthesized within the body itself: vitamins. These substances must be obtained from outside sources such as food and water, or from a man-made equivalent, the familiar vitamin or mineral tablet. Exceptions to this definition include vitamin D, which is made in the body to a limited extent, and vitamin B_{12}, which is produced in the intestinal tract. Although vitamins are found in meats, dairy products, and eggs, they are particularly abundant in vegetables and fruits, an evolutionary gift from our vegetarian forebears. Beta carotene, which converts to vitamin A in the body, is found in dark green leafy vegetables, yellow and orange vegetables, and fruit; vitamin C in citrus fruit, green peppers, strawberries, raw cabbage, and green leafy vegetables; vitamin E in nuts, seeds, whole grains, and vegetables; folic acid in green leafy vegetables; and vitamin K in leafy vegetables and corn.[6]

Many nutritionists see these traditional sources as being far more important than man-made supplements; they distrust any reliance on an artificial source that might lead to bad dietary habits. Until recently, all university-educated nutritionists were taught to scorn supplements, and as one of them notes, "Rare was the nutrition textbook that didn't assail supplement use and equate it with 'food faddism.'"[7] Instead of supplements, these nutritionists pointed to the government's dietary guidelines that suggested three to five daily servings of vegetables, two to four of fruit, six to eleven of grains, and two to three of meat, eggs, poultry, and dried beans. Stick to a diet like that and throw away your tablets, they said, and the FDA agreed with them.

But in the 1980s the realities of modern American life had begun to erode that position, as figures released by the National Center for Health Statistics showed that less then 10 percent of American adults were managing to maintain the government's ideal diet. Equally important, most Americans no longer lived close to where their food was grown, and vitamin levels tend to decline between the farm and the dinner table. The decline is caused by exposure to oxygen, heat, light, and water, with some of the loss occurring during transportation, and some during the preparation of the food. Depending on the means of preparation, as much as 60 percent of vitamin A may be lost in cooking, 100 percent of vitamin C, and similar amounts in many of the other categories.[8] Add to that the indifferent American taste for vegetables, the addiction to fast-food grazing by the young, and the research findings that many pregnant women, senior citizens, invalids, and alcoholics simply cannot obtain the proper nutrients through food alone, and the case for supplements began to be compelling.

A case in point was the announcement by the Public Health Service in 1992, after decades of indifference to vitamins, that all women capable of becoming pregnant should consume 0.4 mg of folic acid a day to reduce the chances of having a child with spina bifida, or other neural tube defects.[9] The PHS did not specify how the folic acid should be consumed, through food or tablets, but at the same time the Texas Department of Health announced a program to distribute free folic acid tablets to low-income women for the same purpose.[10]

With actions such as the distribution of folic acid tablets, nutritionists agree that we have now moved into a second stage in our use of vitamins. The first stage began in the nineteenth century when we first began to understand the role of vitamins in combating nutritional deficiencies such as rickets and beri-beri. In a classic example of those times, the British Navy recognized the value of vitamin C by issuing limes to its sailors as a guard against scurvy on long voyages, thus coining the nickname "limey."

But scurvy and rickets are, for the most part, diseases of the past, and the second stage of vitamin research has shown that nutritional supplements have a marked effect on the general health, and in the prevention of certain chronic diseases. Cancer, neurodegeneration, and heart disease are only some of the disorders that new studies indicate may be forestalled, or even reversed, by one or more of the vitamins in the spectrum that ranges from A to K, and all the subsets in between. Although many of these theories remain to be proven, it is now thought that folic acid can lower the risk of cervical cancer, that vitamin B_6 appears to enhance the immune response in older people, and that vitamin K seems to help maintain bone mass and prevent osteoporosis in postmenopausal women.[11]

Of particular interest are the antioxidants, vitamins C, E, and beta carotene, which may deter the development of cataracts, and have a far-ranging effect on the human body including a slowdown of the aging process. These nutrients appear to intercept those toxic compounds in the body known as oxygen-free radicals, which are created by exposure to tobacco smoke, sunlight, X-rays, and environmental pollutants, as well as by normal metabolism. It has long been thought that free radicals play a role in the development of cancer by damaging cells and their genetic material, DNA (deoxyribonucleic acid), but the antioxidants seem to block this effect by neutralizing the radicals and putting them out of action.

Scientists also believe that the constant assault on the body's cells by free radicals is a factor in the process of aging, and although no one is promising a latter-day fountain of youth, antioxidant therapy may one day be a part of the life-extension process. The evidence uncovered so far by researchers at highly reputable institutions points to a future in which nutrients will play an essential role in human health care, a goal not even dreamed of a dozen years ago.[12]

But even as this evidence began to mount, the FDA continued to view with great skepticism the purveyors of high-fiber products, whole-grain baked goods, and products containing vitamin C, vitamin E, B vitamins, and other recognized nutrients. Reflecting the views of the mainstream medical establishment, the FDA saw the nutritional approach to preventative medicine as something close to disreputable, pointing out that our knowledge of how vitamins and minerals work in the body is still in an investigatory stage, and that some compounds, vitamins A and D in particular, can be toxic if taken in doses that significantly exceed the federal government's recommended daily intake.

By the end of the decade the points of contention between the vitamin libertarians and the FDA had become particularly bitter, as when the Nutritional Health Alliance, an industry group, accused the agency of hiring a hundred new criminal investigators specifically to prosecute vitamin users.[13] Not so, said a high agency official, implying that the FDA had bigger fish to fry. "Those investigators will be dealing with major counterfeiting cases and the drug diversion cases where the agency will be working closely with the DEA [Drug Enforcement Administration] and the FBI."[14]

NHA President Gerald Kessler was unimpressed. "When someone who has been harassing you goes out and buys an assault rifle, you have to figure that he's going to use it on you eventually."[15]

The reference to an assault rifle may have been excessive, but it was also understandable in light of the FDA's various attempts over the years to designate vitamins and minerals as either "drugs" or as "food additives," and thus to regulate the diet supplement industry.

The opening gun was fired on June 20, 1962, when the FDA proposed that all dietary supplement labels include only those nutrients recognized by competent authorities (read FDA-sponsored panels) as being essential and of significant value in human nutrition. The proposals were quickly criticized on the grounds that they would unduly interfere with the rights of consumers, and in the face of such criticism, they were withdrawn.

The FDA, however, did not abandon the idea of revising the supplement regulations. On June 18, 1966, the agency published new regulations pertaining to the labeling and content of "special dietary food products," along with a new definition for vitamin and mineral supplements. Once again, reaction to the agency's proposals was immediate and critical, and because of the strong outcry the FDA was forced to publish an order staying the effective date of the regulations.

On August 2, 1973, the FDA fired its heaviest salvo yet when it published regulations intended to take effect on January 1, 1975.

These draconian measures proposed that most vitamins and minerals with a potency exceeding 150 percent of the U.S. recommended daily allowance be classified as over-the-counter (OTC) drugs. Vitamin A was to be classified as a drug in dosages exceeding 200 percent of the RDA, and vitamin D was to be similarly classified in dosages of over 100 percent of the RDA. These two products would then be subject to prescription drug regulations, not the looser OTC rules, just as if they were antibiotics or opiates.

Again, these proposals drew critical reactions, many of them based on the agency's use of the RDA for vitamins and minerals as a standard, a guideline that even in the mid-1970s was more than thirty years out of date. The RDA had its origins in the early days of World War II when the government established a nutritional standard based on the needs of active teenage boys. The intent was to raise the nutritional level of young men coming out of the hard times of the Depression and into the armed forces, many of them suffering from malnutrition and scurvy. Other RDAs were later established according to age and sex, but since 1968 the FDA had been using those antiquated standards meant for fast-growing teenage boys as the national norm.[16]

Public reaction to these new FDA proposals was so strong that between January 3, 1973, and October 3, 1973, some seventy bills limiting the FDA's authority to promulgate such regulations were introduced into the House of Representatives. These bills were codified and combined, and eventually led in 1976 to the enactment of the Proxmire Amendment to the Food, Drug, and Cosmetics Act.[17] This amendment limited the FDA's authority to regulate dietary supplements, and acknowledged that:

1) The FDA had gone too far in its condemnation of the dietary supplement industry and in its restriction of the public's right to choose its source of nutrients.

2) Dietary supplements should not be treated either as drugs or as food additives.

3) The public was entitled to receive messages about health in conjunction with the sale of dietary supplements.

Passage of the Proxmire Amendment, however, did not stop the FDA from attempting to eliminate dietary supplements in other ways. The agency continued to bring actions in law claiming that supplements, even though they bore no claims for treatment or the prevention of disease, were nonetheless drugs because they were toxic. These cases all failed, rejected by the courts, and it wasn't until 1992 that the agency finally achieved a partial control over the supplement industry.

The FDA gained that control through the use of the Nutrition Labeling and Education Act (NLEA), a much-needed piece of legislation that was designed to curb misleading health claims on all food products. The law went into effect for foods in general on December 31, 1992, but an adroit piece of political horse-trading by Senator Orrin Hatch, whose home state of Utah hosts a $700 million per year vitamin industry, gained a one-year moratorium for the makers of nutritional supplements. Once that moratorium expired, however, the manufacturers of food supplements would be allowed to make health claims for their products only if those claims were to be supported by "significant agreement" within the scientific community. If this agreement was not forthcoming, and if the health claim was not approved by the FDA, then the manufacturer would not be allowed to use it.[18]

On the surface, this appeared to be a reasonable requirement, but the libertarians saw it as a threat to health-care freedom of choice and their right to choose beneficial nutritional supplements. And their point of view had a certain justification, since the scientific community that must provide the "significant agreement" is the mainstream medical establishment, which is only now beginning to recognize the therapeutic values of vitamins and minerals; and they feel that the FDA, after decades of antisupplement bias, cannot be trusted to adopt an even-handed attitude.

But by the end of the 1980s, the attitude at the FDA had undergone a subtle shift of emphasis. It would have been too much to expect the agency to make a complete about-face after decades of antisupplement bias, but the agency's senior advisor to the commissioner, Mary Pendergast, was able to admit that:

> It's fair to say that the FDA once had a reputation for pooh-poohing the idea of taking vitamins, but that attitude is changing. We no longer disagree with anyone's desire to supplement their diet with a moderate dose of vitamins and minerals. We do worry, however, when products are not properly labeled, and when false and fraudulent claims are made. We also worry about people taking megadoses, because that can be harmful. Past that, we don't care. If people want to supplement their diets, that's fine with us.[19]

But the supplement industry, and those who support it, see no change at all in the FDA position. The agency is still the enemy, and the industry has placed its hope for regulatory relief legislation on two bills that, at this writing, are due to be introduced in Congress. One is authored by Senator Hatch, and the other by Representative Bill Richardson (D–N.M.). Both lawmakers have geared their bills to industry charges that the FDA arbitrarily seizes products in the name of safety and that it has proposed standards for health claims on labels that are unrealistic and unreachable. The bills would restore control over health claims to the manufacturers, requiring only that they label their products with truthful information that is not misleading. No FDA approval of the labels would be required.[20]

The scare word in the field of supplements is megadosing, but what exactly is a megadose? The word provokes an image of laymen irresponsibly popping pills by the fistful, but when practiced by physicians under the name of orthomolecular medicine, it takes on an air of respectability. Nobel Laureate Linus Pauling was the first to use the term to describe the treatment of disease with nutrients that occur naturally in the body. His basic premise was that the greatest long-term health benefits could be derived through the concentration of essential nutrients by diet adjustment and the ingestion of massive doses of various vitamins and minerals.[21] Initially used to treat schizophrenia, orthomolecular principles have been used over the years to treat epilepsy, autism, senility, arthritis, and various allergies. But always under the care of a physician, since the risk of overdosing is ever present. A doctor might well prescribe a dose of 2,000 mg of niacin, which is 100 times the maximum amount recommended, in an attempt to lower a patient's cholesterol level, but such a patient must be constantly monitored for signs of jaundice and liver damage, common side effects of too much niacin.[22]

Overdosing on vitamin A can also lead to liver damage, hair loss, and blurred vision, while too much vitamin D can interfere with muscle functions, including that of the heart. Even when conducted by qualified physicians, megadosing is a controversial procedure that is used by relatively few doctors. When used by an unqualified person, it comes dangerously close to self-medication. In the case of niacin, which is related to the nicotine in tobacco, the Council for Responsible Nutrition has urged its member companies to adopt a maximum level of 500 mg, and to carry a cautionary label that suggests the supervision of a physician. CRN has also issued maximum dosage recommendations for vitamins A and B_6.[23] Still, many people continue to ingest large amounts of vitamins C and E, spiced by a calcium pill or two, and with a multipurpose capsule to top it all off like a cherry on an ice cream sundae. Shopping for vitamins and minerals is as easy as shopping for groceries. It's all there, right on the shelves.

The easy availability of dietary supplements is based on the assumption that they start out in life as foods, not drugs, and therefore are not subject to strict regulation. But when does a food become a drug?

"This is one of the most difficult interfaces that we have in food and drug law," says Pendergast. "There was a time when we found honey being sold as a cancer cure. Honey, of course, is a food, but when you make a claim like that it has to be treated as a drug. And if the claim is made as a drug, then the manufacturer has to have two adequate and well-controlled investigations showing that the product lives up to its claims. Just like any other drug."[24]

But when a claim is purely nutritional, then the product is treated as a food, as when claims are made that calcium helps to prevent osteoporosis, or that fiber helps to prevent colon cancer. Then the drug rules do not apply, and all that need be shown is the "significant scientific agreement," that the claim is valid. When it comes to dietary supplements, the line between food and drug is a thin one, and the distinction does not sit well with the vitamin community, which would prefer to see all supplements treated as foods.

Privately, many both in the FDA and in the supplement industry agree that what is needed are realistic guidelines to define the optimal consumption of dietary supplements based on age, sex, and lifestyle. What is also needed is a relationship between the FDA and the supplement industry that is less confrontational and less dependent on outmoded attitudes. A step in that direction was taken in September 1992, when the FDA's deputy commissioner for policy, Michael R. Taylor, extended an olive branch to industry at the annual conference of the Council for Responsible Nutrition. Conceding that the traditional FDA position had been that nutrition should come only from a balanced diet, he now agreed that, "Many Americans don't, in fact, get all the nutrients they need from their diets, and . . . nutritional supplementation is, for them, at least a prudent precaution, if not an absolute necessity for good health."[25]

The statement was a small, but important, step toward conciliation. Nobody expected the FDA to endorse the sort of megadosing that was common among serious vitamin takers, but then came the news, in May 1993, of two studies of more than 120,000 men and women, which strongly suggested that supplements of vitamin E could significantly reduce the risk of disease and death from fat-clogged coronary arteries. The studies, by researchers at the Harvard School of Public Health and at Brigham and Women's Hospital in Boston, showed that initially healthy people with the highest intakes of vitamin E developed coronary disease at a rate of about 40 percent lower than comparable men and women whose intake of the vitamin was lowest. This occurred independently of any change in the blood levels of cholesterol.[26]

The greatest protection was found at levels of about 100 daily international units of vitamin E, with its high antioxidant value, while the federal recommended daily allowance was only 15 units. This was a clear case of worthwhile megadosing. The findings appeared in the *New England Journal of Medicine*, and were some of the first to find health benefits from extreme doses of vitamins. Still, a formal health claim would have to wait until stringently designed clinical tests had established both the benefits and the risks of vitamin E supplements.[27]

The news came as no surprise to vitamin enthusiasts who managed, at least publicly, not to thumb their noses and say, "I told you so." Many researchers in the field now admitted privately that they and their families had been taking high doses of the antioxidants long before there had been enough hard evidence to justify recommending the practice to the public. Now they suggested an open-minded and inquisitive approach to the entire subject of dietary supplements. Eat a well-balanced diet moderately supplemented with essential vitamins and minerals, they said, but keep informed, read the literature, and be prepared for more unfolding discoveries. The serious vitamin takers may turn out to be mere eccentrics, but, despite an institutionalized, decades-long opposition by the FDA, they also may turn out to be the nutritional prophets of the age.

The agency at its best went virtually unnoticed when, on March 28, 1979, the Three Mile Island nuclear reactor near Harrisburg, Pennsylvania, came perilously close to a complete meltdown. Within hours, FDA field inspectors were out taking samples of milk, fish, and water within a 20-mile radius of the facility. During the first twelve days of the emergency, the inspectors collected more than five hundred samples from dairy farms, rivers, food processing plants, and retail stores. Laboratory tests showed that there was no threat to the public health, but the potential had been there for a disaster and the FDA had been on the job.[28] There was no mention of FDA activity in any of the communications media.

On April 26, 1986, the Soviet Union's nuclear facility at Chernobyl exploded. When the FDA learned of the disaster, it placed more than a thousand agency employees on an alert status to check imported foods that might have been contaminated abroad and domestic foods that might have been contaminated when the radiation cloud reached North America. Again, the increase in radiation levels proved to be minimal, but again the potential for disaster had been there.[29] And again, the public was never informed of the FDA's role.

In 1989, the San Francisco earthquake, the wreck of the Exxon Valdez, and Hurricane Hugo combined to give the FDA field crews a triple exercise in disaster response. In the wake of Hugo's destructive pounding of South Carolina, Puerto Rico, and the U.S. Virgin Islands, teams of FDA investigators were sent in to help in the cleanup as well as to ensure the safety of the food and drug supplies. Millions of dollars worth of contaminated products were condemned, the public health was protected, and once again the good news was no news at all.[30]

Thus, if there was one area during the 1980s that the FDA could point to with pride, it was the agency response to natural and man-made disasters. These, for the most part, were shining moments for the FDA and its division of emergency and epidemiological operations. At times it seemed as if only during these moments of crisis could the agency summon up the dash and the ingenuity that was so sadly lacking in its day-to-day affairs. Of equal sadness was the absence of any press coverage of these events. No one seemed to care whenever the FDA was on the side of the angels.

But there was no lack of journalistic attention to a different set of crises, the tampering scares involving both food and drugs that the agency faced during the decade. In 1982 it was Tylenol capsules laced with cyanide, and in 1984 it was fragments of glass in Gerber fruit juice products. Gerber was hit again in 1986, as was Tylenol. Excedrin was attacked, and there were dozens of other tampering threats that never made major headlines. Some of those threats were out-and-out hoaxes, some the results of natural causes (the tiny bits of Gerber glass were part of the bottling process),[31] but some were acts of industrial terrorism that took lives and sent profound shock waves across the country.

It was the FDA that had to deal with those crises, and the agency dealt with them well, issuing timely warnings and working closely with industry to pull products off the shelves. There were relatively few deaths, and if the financial damage ran high, the victims were healthy corporate giants that were able to rebound well.[32] But within the FDA each crisis produced its own strains, its own anxieties, its own contributions to the institutional mindset, and somewhere along the line the fear of product tampering, particularly by cyanide poisoning, became the full-time bogeyman of the upper echelons of the agency. Everybody was waiting for "the big one" to happen, a major eruption of terrorism by tampering that would hold the entire nation hostage. No one knew what form it would take, or whether the motivation would be financial, ideological, or mindlessly maniacal, but everyone was sure that it was on the way.

The big one finally arrived in the form of a couple of Red Flame grapes from Chile that frightened the socks off the American consumer, swept tons of fruit off supermarket shelves, and severely damaged the Chilean economy. The Chilean Grape Scare of 1989, which was first perceived as an averted tragedy, and then as a severe overreaction by the Food and Drug Administration, eventually proved to be nothing less than a farce. After it was all over a consortium of Chilean fruit growers brought a lawsuit against the U.S. government to recover more than $330 million in damages caused by the FDA's eleven-day embargo of their products, and the case contained a startling accusation. According to the growers, the suspect grapes that caused all the fuss were most likely contaminated not by terrorists in Chile, as first was thought, but in the FDA's own Philadelphia laboratories.[33] One scenario had it that the grapes were accidentally contaminated while being tested for cyanide, while a second theory suggested that the FDA had conspired with the U.S. State Department to poison the grapes as an act of economic sabotage against the regime of the then-Chilean leader, General Augusto Pinochet.

Other subplots abounded, involving Chilean communists, a hit team of Israeli fruit mavens, and a California grape cartel, but the growers were wise enough to leave the conspiracy bones for the buffs to chew on. They did insist, however, that the grapes had been clean until they came into the hands of the FDA. The growers made a compelling case for accidental contamination, and the charges were a distinct embarrassment to an already embattled agency. As the Washington wags had it, it is one thing to shoot yourself in the foot, and something else again to poison your own grapes.

Both sides agreed on the bare bones of the story. On March 2, 1989, an anonymous caller to the American embassy in Santiago warned that fruit headed for the United States had been injected with cyanide as an act of protest against the Pinochet regime.[34] The call was repeated on March 8,[35] and two days later FDA Inspector William Fidurski spotted three suspicious-looking grapes on the docks at Philadelphia, part of a shipment that had just come in from Chile. Two of the grapes had puncture marks, each mark surrounded by concentric white rings; the third grape was simply slashed. The fruit was rushed to the FDA Philadelphia laboratory, where the two punctured grapes were tested and found to contain cyanide.[36] Not very much cyanide, not even enough to make a small child feel ill, but then–FDA Commissioner Frank Young announced that it was better to be safe than sorry, and the race was on to see who could dump the most Chilean fruit.[37] (That wasn't the only race. In Oregon, a distraught mother sent a police car on a high-speed chase after a school bus when she suddenly remembered that she had packed some grapes in her daughter's lunchbox.)[38]

The FDA impounded 2 million crates of fruit at airports and docks across the country, and advised consumers not to eat any fruit from Chile, which included most of the peaches, blueberries, blackberries, melons, green apples, pears, and plums that were on the market at that time of year.[39] Japan and Canada quickly followed suit, and before the dust had settled, 20,000 Chilean food workers had lost their jobs, hundreds of thousands more had suffered economic distress,[40] and a little girl in Oregon was still trying to figure out why that cop had commandeered her lunch. The dust settled with a whimper after an eleven-day orgy of dumping when the FDA announced that no other poisoned produce had been found, and that Chilean fruit was really rather good for you.[41]

Staring at the bare bones it would seem that, at the very worst, the FDA had erred only on the side of extreme caution, but an entirely different story began to emerge once those bones were covered with flesh.

On March 11, 1989, the SS *Almeria Star,* thirteen days out of Valparaiso, Chile, with a load of fruit, arrived in the port of Philadelphia, and discharged its cargo at the Tioga Marine Terminal. It was now nine days since the first anonymous telephone call to the embassy in Santiago, and they had been days of intense activity at the FDA. The first move, taken in conjunction with Customs authorities, had been to detain all Chilean fruit while the situation was evaluated and studies were performed at the FDA's emergency operations laboratory in Rockville, Maryland. The studies, conducted over the weekend of March 4 and 5, resulted in two conclusions.[42]

The first noted that the caller had specifically used the word *inject,* and the studies confirmed that the internal pressure of the fruit pulp in grapes would force virtually all of any cyanide injected back out onto the surface of the grape. This would cause the fruit to shrivel and turn black, making it easily distinguishable. It was hardly an effective terrorist technique. In addition, the studies confirmed that cyanide injected into a grape would rapidly dissipate to low levels soon after injection, a reaction to the sugars and acids naturally present in the fruit. There was no way in which a hypodermic needle could be effectively used to inject a hazardous dose of cyanide into a grape.[43]

The second conclusion was that, in the absence of any credible means of carrying out the threat, the anonymous warning was probably a hoax.[44] Despite this, and just to be safe, the FDA would mount an intensified inspection procedure for all fruit coming into the country from Chile.

This was the FDA at its best. The agency had met the situation head-on, had scientifically evaluated the risks involved, and had come to the logical conclusion. Unfortunately, this was the high point of the FDA's involvement in the affair. Once the crunch came, all the science would go out the window, logic would give way to panic, and Chicken Little would go racing round the barnyard. In the end, the agency would decline to believe in its own research, and, for that matter, in itself.

The three suspicious-looking grapes found by Inspector Fidurski, two punctured and one slashed, were sent to the Philadelphia laboratory for testing, and as a matter of routine, along with them went the remaining grapes in the shipping box, the box itself, the wrapping material, and eight other boxes of grapes from the same shipping pallet.[45] At the laboratory, the two punctured grapes were macerated into pulp and half of the resulting sample was subjected to two procedures designed to show the presence or absence of cyanide. The cyantesmo strip test indicated the presence of cyanide, and the chloramine-T test indicated the level of contamination. Both tests registered positive, and the operations center in Rockville was advised that poisoned grapes had been found.[46]

But while the Philadelphia chemists were congratulating each other on their discovery, the other half of the pulp sample, along with the slashed grape and the rest of the grapes in the bunch, were all on their way to the FDA laboratory in Cincinnati for confirming tests by Dr. Fred Fricke, an expert in cyanide contaminations. It was just after midnight when Cincinnati advised Rockville that they had tested their half of the pulp sample, as well as the entire slashed grape. They had also tested all of the other grapes that had been sent to them, including two more that bore the same concentric white rings that had aroused suspicion in the first place. Further, they had tested the wrapping tissue, the wrapping plastic, and all the other packing material that had accompanied the grapes.[47]

And they had found nothing. Using the same tests as the chemists in Philadelphia, they had not been able to find a single trace of cyanide contamination. Not in their portion of the pulp, not in the other grapes, not in the packing materials. Nothing.[48]

The news presented Commissioner Young with a disquieting set of options. Two FDA laboratories, using the same methods, had come up with diametrically opposing results. Obviously, the next step was to retest the Philadelphia sample, but along with the Cincinnati results came the unhappy news that the Philadelphia sample had been—oops—destroyed during the testing process. At one point in the analysis it had been necessary to neutralize the sample before proceding. There were two ways of doing this, either by freezing with liquid nitrogen, which would preserve the sample, or by the addition of sodium hydroxide, which would trigger an irreversible chemical reaction that would eventually destroy the sample.[49] Inexplicably, the Philadelphia laboratory had chosen the sodium hydroxide method in clear violation of both the *Code of Federal Regulations* and the *FDA Regulatory Procedures Manual,* which require that part of any original sample must be kept aside for testing by the owner of the product or the owner's attorney. Whatever the reason for this decision, the Philadelphia half of the sample was now destroyed and unavailable for further testing. Of course, there was always the other half of the sample in Cincinnati, but that had already been tested as negative.

It was now 3:30 in the morning of March 13, and a decision had to be made. Young had the following points to consider.

Well before the discovery of the suspect grapes, his own experts had told him that it would not be possible to poison a Chilean grape by injection with cyanide.[50]

The Philadelphia laboratory had reported a level of cyanide in its sample so low that it would not harm a small child.[51]

The Cincinnati laboratory had found no sign of cyanide either in its half of the sample, or in any of the other material connected with the case.[52]

Based on the Cincinnati analysis, there was no connection between the concentric white rings and cyanide contamination. Cyanide in the grapes would have backed up onto the surface and would have caused the fruit to shrivel and turn black.[53]

Cyanide is a migratory substance—it travels easily. If the two suspected grapes had actually been injected with cyanide, then the surrounding grapes in the bunch and in the box should also have shown traces of the poison. Yet they did not.[54]

The box that had contained the suspect grapes had been placed in such a position aboard the *Almeria Star* that it was inaccessible during the entire voyage from Valparaiso to Philadelphia. That left only Chile and the United States as possible venues for tampering.[55]

The first of the tests applied in Philadelphia consisted of exposing a strip of reactive paper to gases issuing from the sample. In the presence of cyanide, the paper would turn blue; the more cyanide, the bluer. Philadelphia had reported a deep blue color,[56] but that only proved that the grapes could not have been tampered with in Chile, for after thirteen days at sea the paper should not have reacted at all.

In the same sense, if the tampering had taken place in Chile, the grapes would have looked black and shriveled after thirteen days on the *Almeria Star*.[57]

Since the suspect grapes had first been discovered at the Tioga Marine Terminal, they had been under the direct control of the FDA.

And then came the clincher, something that Young knew, should have known, or should have been told. The proper execution of testing to determine cyanide contamination requires the presence of cyanide in the testing laboratory. Specifically, cyanide is used in such cases to calibrate the testing instruments. It is also used to "spike" the sample being tested, which means to introduce a known amount of cyanide into the sample to guard against false negative results. On March 12, 1989, cyanide was used for just those purposes in the Philadelphia laboratory of the FDA.[58]

It was most probable that the grapes had been contaminated during that process.

Those were the factors that Young had to work with, and it would seem that his course was clear. All the evidence at hand indicated that the warnings in Santiago had been part of a hoax, that there had been no tampering, and that the grapes had been contaminated in the Philadelphia laboratory. Despite this, he proceded to announce the embargo of all Chilean fruit coming into the country, and later that day, appearing on national television, he urged consumers to dispose of all the Chilean fruit they might have on hand and instructed merchants to withdraw such fruit from their shelves.

Better safe than sorry was the rationale, but was it only that? Was it, instead, a reluctance to admit that the Philadelphia laboratory had dropped the ball, combined with the institutional mindset that existed within the FDA at that time? Had the agency seen those two lonesome grapes as the start of "the big one," the ultimate act of international terrorist tampering that had been predicted and expected for years? Apparently, that was the way that Frank Young saw it. To be kind, one would have to assume that, despite all the evidence to the contrary, he really believed it. Referring back to the second Tylenol poisoning, he said, "We were prepared intellectually for this day since 1986. I was waiting for this to happen. This was not at all a surprise. [I knew] it would eventually come to a major international tampering."[59]

Having conceived the event, could the agency do anything less than give it birth?

When the embargo was announced, nothing was said about the negative results of the Cincinnati tests, nor about any of the other factors that made such a strong case against tampering. Nothing was said about where the contamination probably had occurred. The public was told only that poisoned grapes had been found, that a clear and present danger existed, and that the sky was falling down. The "big one" had finally arrived; Chicken Little screamed and a nation trembled.

In the end, the Great Chilean Grape Scare would have been hilarious, had it not been so sad. It had little to do with public health, for the public health was never in danger. It had little to do with Commissioner Young; he was gone from the agency within the year. It had much to do with money, but that was a matter for the courts. It had much to do with terrorism in our time, and how we have been conditioned to respond to it.

But more than anything else, it had to do with the manner in which the FDA functions in our society. Risk assessment lies at the center of what the agency does every day. Should a new drug be allowed on the market? Should a food product be banned because of an ingredient linked to cancer? How much of a medication is enough, and how much is too much?

These judgments are often quantitative in nature, and a risk that is rated at one in a million has to be treated with less concern than a risk that is rated at one in ten thousand. If such distinctions are not made, then the decision-making process is flawed, and cannot be relied upon, whatever the odds and whatever the situation. It is no defense to mouth the obvious "better safe than sorry." The Chilean Grape Scare showed that one can be safe and look sorry at the same time.

7

The AIDS
Gridlock

THE STORY OF THE SEARCH FOR DRUGS AGAINST AIDS (ACQUIRED immunodeficiency syndrome) in the 1980s has been well chronicled. More than a few good books have been written about the indifferent lack of leadership on the part of the Reagan and Bush administrations, about the hidebound federal bureaucracies that often did more harm than good, about the giddy rush for profits by the pharmaceutical companies, about the often controversial tactics of the AIDS activists. The titles of those books can be found in the Notes at the end of this book, and it is not the intent here to go over the ground they have covered so well. The brief of this book is restricted to the actions, and lack of them, taken by the FDA during that time, and it is in a sense a story with a reasonably happy ending, for the agency response to the epidemic began with apathy and sloth and ended with a realistic and workable modification of the agency's entire drug approval program. Indeed, modification may be too mild a word, for what happened in the 1980s was a sea change not only in FDA procedure but in FDA philosophy as well.

Little of this would have come about if the agency had been left to its own devices. During the 1980s, getting any sort of action out of the FDA that did not conform to traditional procedures was like getting a Missouri mule to dance the two-step. First you had to hit it over the head with a two-by-four just to get its attention. Then you had to teach it how to dance. The clout on the head and the subsequent instructions in footwork were all supplied by the various AIDS advocacy communities in an activist campaign that was unique in the experience of the agency. Concerted, skillful, and often outrageous, it was a campaign that forced the FDA to take a long, hard look at itself, to confront its faults, and to do something about them. This was a monumental accomplishment, one that would have been judged impossible before the fact by those who knew the agency best. Fueled by indignation and confronted with elemental choices, the AIDS community accomplished the impossible by battling the FDA at every level from the streets to the boardroom, insisting on changes in the way that drugs were made available to the American public, and by the end of the decade the drug approval process in this country had been radically altered.

The drug approval procedure, as we have seen, was rooted in the Kefauver Amendments of 1962 that required the pharmaceutical manufacturer to prove that not only was its product safe to use but that it actually did the job for which it was intended. This new concentration on the efficacy of the product soon led to charges that the agency, under the domination of the pharmaceutical industry, was now remiss in protecting the public against unsafe drugs. In August 1974, Senator Edward Kennedy conducted a series of hearings into these charges, and the agency responded bitterly. Denying any negligence, Commissioner Schmidt testified that rather than being unduly influenced by industry, by far the greatest pressure on the FDA came from congressional hearings designed to discourage the approval of new drugs. No hearing had ever been conducted, said Schmidt, to investigate the failure of the FDA to approve a new drug, and countless hearings had been conducted to criticize the approval of one.[1]

Leaving aside Schmidt's questionable denial of industry influence, his response pointed up a division of priorities in the FDA approval structure that Grabowski and Vernon later illustrated in their work on the regulation of pharmaceuticals.

		New drug is safe and effective	*New drug is not safe and effective*
	ACCEPT	Correct decision	Type 2 error
FDA decision			
	REJECT	Type 1 error	Correct decision

By their definitions, if a new drug is safe and effective, then an FDA decision to accept it is considered correct, and an FDA decision to reject it is classified as a type 1 error. Conversely, if a new drug is not safe and effective, then an FDA decision to reject it is considered correct, and a decision to accept it is classified as a type 2 error.[2]

On the surface it would appear that the two types of errors carry equal weight, but Grabowski and Vernon point out that:

> It can be plausibly argued . . . that our regulatory structure does not have a neutral stance between type 1 and type 2 errors. The mandate to the FDA is drawn in very narrow terms . . . to protect consumers against unsafe or ineffective drugs [that is, to avoid type 2 errors]. There is no corresponding mandate to avoid type 1 errors, or to compel equal concern with new drug innovation and improved medical therapy. In point of fact, the institutional incentives confronting FDA officials strongly reinforce the tendency to avoid type 2 errors at the expense of type 1 errors. The cost of rejecting a good drug is borne by outside parties [manufacturers and patients], while the cost of accepting a bad drug is borne by the official and the FDA as a whole.[3]

Beginning in 1962, when the efficacy of drugs became part of the FDA mandate, this inbuilt desirability to avoid type 2 errors colored the thinking of the agency, and of its entire approval process. But the Kefauver Amendments would never have become law without the public outcry over thalidomide, and, as Edgar and Rothman tell us:

> The entire episode demonstrates how powerful the symbolic role of a nightmare case can be in the implementation of public policy. Sustaining the drug regulatory enterprise between 1962 and the AIDS crisis was the figure of an heroic Frances Kelsey, single-handedly

saving Americans from tragedy by saying no to a drug manufacturer. The message was clear: those who exercise caution reap rewards. . . . Moreover, this message was one to which the FDA staff was especially receptive, for those recruited to these positions, at salaries substantially below those in the private sector, were very likely to arrive with a sense of mission about consumer protection.[4]

Better that one unsafe drug be kept from the marketplace than that ten safe medicines be presented to the public. There was, after all, a plethora of medicines already available for almost any human ailment, with more being presented for consideration every year; and for the needs of those with rare illnesses, there was the Orphan Drug Act, which gave tax incentives to those manufacturers who were willing to develop drugs for those markets of fewer than 200,000 potential users.

But despite this institutionalized conservatism, bad drugs did, and do, get to the public:

Omniflox, a broad-spectrum antibiotic, had to be recalled after less than six months on the market due to reports of severe adverse reactions including three deaths.[5]

Within eighteen months after Versed, a sedative, went on the market in the United States, the FDA received reports of eighty-six adverse reactions including forty-six deaths.[6]

After years of use, fenoterol, an inhaled drug designed to relieve asthma attacks, was found actually to increase the risk of death rather than lower it.[7]

Oraflex, an antiinflammatory drug that was hailed as a breakthrough for the 30 million Americans who suffer from arthritis, had to be recalled within three months of approval after reports that the medication had caused the deaths of seventy-two users in the United States and Great Britain.[8]

Despite these slips through the regulatory cracks in the floor, the concentration of the agency from the 1960s onward was exclusively concerned with the prevention of type 2 errors, and it is against that background that one must appraise the conduct of the FDA when the AIDS crisis came upon the nation and the world.

As thousands of people died, the agency's slow and deliberate approval process suddenly appeared to be both inappropriate and perverse. People with AIDS first begged, and then demanded, that the government allow untested drugs, and those drugs still in the testing process, to be made available at once. To the FDA, this appeared tantamount to tearing down the consumer protection laws that had been built up over two decades, and appeared to invite a spate of the type 2 errors that the agency had worked so hard to avoid. This was unthinkable to the agency mindset, and so, along with the National Institutes of Health, the Centers for Disease Control, and the National Institute of Allergy and Infectious Diseases, the FDA insisted that those who were dying would simply have to wait for science to produce a cure.

Despite this passivity, which, at least, could be defended as scientific caution, active intervention at the highest level might have modified the FDA position, but no such action was forthcoming. As Arno and Feiden point out, "For the first four years of the epidemic, Ronald Reagan refused to utter the word AIDS. Conservatives in charge at the White House were so offended by gay lifestyles and sexual practices that a trickle of AIDS cases was allowed to become a flood before any action was taken. Racism and disdain for drug users, who are mainly poor and often black and Hispanic, intensified the problem."[9]

The following section of Samuel Butler's nineteenth-century satire *Erewhon* was much cited in the early days of the epidemic as a symptom of the association between AIDS and criminality in the national consciousness, a place where a righteous morality seemed strangely blended with science:

> In that country if a man falls into ill health or catches any disorder or fails bodily in any way before he is seventy years old, he is tried before a jury of his countrymen and if convicted is held up to public

scorn and sentenced more or less severely as the case may be. But if a man forges a check, or sets his house on fire, or robs with violence from the person, or does any such things as are criminal in our own country, he is either taken to a hospital and most carefully tended at the public expense, or if he is in good circumstances he lets it be known to all his friends that he is suffering from a severe fit of immorality, just as we do when we are ill, and they come and visit him with great solicitude—for bad conduct, though considered no less deplorable than illness with ourselves, and as unquestionably indicating something seriously wrong with the individual who misbehaves, is nevertheless held to be the result of either pre-natal or post-natal misfortune.[10]

Some diseases, it seemed, were punishments to be borne with a bowed head as a rebuke from the gods. "For several generations now," Sontag pointed out, "the generic idea of death has been a death from cancer, and a cancer death is experienced as a generic defeat. Now the generic rebuke to life and to hope is AIDS."[11] Altman pointed effectively to the historical position of seeing epidemics as the result of social collapse and degeneracy. Thus, the Black Death was a sign of God's displeasure, and in nineteenth-century America it was cholera that was seen as a consequence of widespread sin. "Certainly," he said, "the response to AIDS seems closer to that described by Butler than to contemporary assumptions about disease."[12]

Sontag further pointed out that:

The sexual transmission of this illness, considered by most people as a calamity one brings on oneself, is judged more harshly than other means—especially since AIDS is understood as a disease not only of sexual excess but of perversity. (I am thinking, of course, of the United States, where people are currrently being told that heterosexual transmission is extremely rare, and unlikely—as if Africa did not exist.) An infectious disease whose principal means of transmission is sexual necessarily puts at greater risk those who are sexually more active—and is easy to view as a punishment for that activity. True of syphilis, this is even truer of AIDS, since not just promiscuity but a specific sexual "practice" regarded as unnatural is named as more endangering.[13]

Thus, with the bigots and the religious fanatics in a position to set the tone of government response to the crisis, there was no room for an aggressive form of leadership that might have prompted the FDA to move even an inch away from the status quo ante. Only after the AIDS community had roared, screamed, and bullied the subject onto the front pages and the six o'clock news did the FDA begin to stir itself into a search for a better way to approve certain drugs. The first of those drugs was AZT (azidothymidine), and its story will be told in the following chapter. These were some of the others.

Pentamidine

A clear indication of the way things were, and how they were likely to remain, came with a drug that was used for AIDS patients who had contracted Pneumocystis carinii pneumonia (PCP). In the early days of the epidemic about 80 percent of those patients with AIDS also came down with PCP, and it was often the first sign that the attacking virus was present in the body. PCP is a particularly brutal form of pneumonia that scars the lungs and debilitates the body; the patient might survive one attack, or two, or three, but no more than that. The organism that causes the disease, *Pneumocystis carinii,* is not limited to people with AIDS. Perfectly healthy people can be infected with it, but the immune systems of perfectly healthy people will suppress the organism and keep it from doing harm. Only when the immune system weakens, and crumbles, does the organism take over, multiply, and kill.

The drug that would eventually be used to combat PCP was called pentamidine, and there was nothing new about it. In the decades before the AIDS epidemic it had been used in cases of hypoglycemia, and to prevent African sleeping sickness. Manufactured in Great Britain, injectable pentamidine was available, but barely. It had never been used in great quantity, the demand for it so slight that in pre-epidemic days it was available in the United States only in small amounts through the Centers for

Disease Control in Atlanta. All that changed with the coming of AIDS. By the end of 1983, the CDC had received more than 2,200 requests for pentamidine, far beyond its ability to supply, and it was clear that a reliable American source for the drug was needed, and quickly.

Finding such a source was anything but easy. CDC officials canvased the pharmaceutical field, looking for a manufacturer that would be willing to produce injectable pentamidine, but as happened over and over again in the early days of the epidemic, most drug makers were unwilling to invest in a product designed for what they perceived to be a highly limited market. Public health officials finally interested an Illinois-based manufacturer named Lyphomed, and cut a deal by which the CDC would pay the manufacturer a flat fee for every vial of injectable pentamidine produced.

The inducement for Lyphomed was the recently passed Orphan Drug Act, which provided substantial tax benefits for the maker of a drug aimed at a market of fewer than 200,000 users. It seemed a good deal all around, and the FDA got into the act by inviting Lyphomed to apply for approval of injectable pentamidine to be used against PCP. There was no question that the approval would be granted. The safety and efficacy of injectable pentamidine had long been demonstrated through published studies, and Lyphomed was thus excused from conducting the usual clinical trials. Six months after application, in October 1984, Lyphomed was granted the exclusive right to sell injectable pentamidine for the seven-year period allowed under the Orphan Drug Act.

There was, however, a problem. Pentamidine, when injected into the body, lodged primarily in the kidneys, spleen, liver, and adrenal glands. Very little of it got to the lungs, which was the prime site of infection for AIDS patients, and almost half of the patients who took the drug developed toxic reactions to it. There had to be a better way to deliver the drug, and researchers quickly came to the conclusion that the best possible route would be through an aerosolized version that could be inhaled directly into the lungs. As Arno and Feiden note, a study had shown that, "When rats were put into a room filled with aerosolized pentamidine, the drug penetrated their lungs in high concentrations while leaving other organs largely unaffected."[14] Aerosolized pentamidine, it was true, had never been tested on humans, but given the speed and ease with which the FDA had granted approval to the injectable version, the people at Lyphomed and the concerned members of the AIDS community felt confident that approval of the new version would only be a matter of time. It turned out to be exactly that—a matter of five long years.

In retrospect, it now seems obvious that Lyphomed's exclusive rights to produce and market injectable pentamidine should have been extended to the aerosolized version, if not automatically, then with genuine dispatch. Desperate people were dying, and they needed the drug, which was known to work. But the consumer-protective mindset that had been in place for twenty years at the FDA, the built-in fear of committing a type 2 error, was stronger than any feelings of fairness and compassion. From the FDA point of view, aerosolized pentamidine was a new drug, wholly removed from the injectable version, and thus subject to the full range of clinical testing that any new drug must endure.

And one cannot entirely fault the agency attitude. The delivery method of a drug—oral, injectable, aerosole—can have a determining effect on the efficacy of the medication, and it is not unreasonable to insist on separate trials for each system. What was unreasonable, however, was to withhold the drug from desperate, dying patients. The times demanded a different response. To people with AIDS, the world was in flames. To the FDA, it was business as usual. Standards were standards, and they had to be met.

For all of that, the situation was still salvageable. Intelligently organized clinical testing, rushed through on a crash basis, could have brought aerosolized pentamidine onto the market in a matter of months. But intelligence was in short supply, and the FDA was not the only government agency with a wrongheaded bent.

Research on the project was placed in the hands of the National Institutes for Allergies and Infectious Diseases (NIAID), a part of NIH, which later spun off the AIDS Clinical Trial Group (ACTG). In addition, the National Cancer Institute also played a role. The result was a stew of committees and subcommittees that, over a period of years, met, argued, proposed, examined, and rejected every nut, bolt, and screw in the process. Arno and Feiden note that the response to the aerosolized pentamidine crisis made it clear that the ACTG research program simply was not working. "NIAID's own subcommittee said in February 1987 that this research ought to be a priority of the highest magnitude. Although some community physicians aerosolized the liquid drug for their patients, no patient was actually enrolled in a government-sponsored study until March 1988."[15]

Nor could a patient be faulted for a reluctance to participate in such a study, for the FDA standards of the time still insisted on double-blind placebo-controlled trials. This meant that half of the patients enrolled in the trial would receive the drug, and the other half would receive a worthless dummy pill. Further, as a double-blind, neither patients nor researchers would be aware of who was receiving what. This made all the scientific sense in the world, but it was nonsense to patients with PCP. Aerosolized pentamidine was available in the rapidly growing underground AIDS drug market, and, as noted, it was available from those physicians who were reconstructing the injectable version. What's more, everyone knew that the drug worked. Why, then, would any sensible person partake in a study where he or she would have only a 50 percent chance of getting the drug?

Time dragged on as the various government entities fenced with each other, made new rules and changed them daily, and fought the battles of proper dosage and preferred delivery systems. In December 1987, it seemed that some progress was being made when three clinical trials, one for PCP treatment and two for prevention, were finally scheduled, and in the following March one of those trials actually got started. But a few weeks later the FDA put a stop to the trial and requested changes in the study design of the other two. There were still a few T's to be crossed and I's to be dotted. When NIAID at last began to enroll patients for their trials in June 1988, physicians across the country and AIDS community activists were well advanced into studies of their own. One year later, Lyphomed received FDA approval to market aerosolized pentamidine, based largely on data developed by these community-based trials.

Ganciclovir

Ganciclovir was another drug that worked. It did the job for which it was intended. Everybody knew that it worked—everybody in the private and the public sector. The AIDS patients knew it, physicians knew it, and its manufacturer knew it. The NIH knew it, the NIAID knew it, and the health commissioners of every major city in the country knew it. Even the FDA knew it. But ganciclovir spent six years on the shelf before the FDA finally approved its use.

Ganciclovir was a drug targeted against Cytomegalovirus retinitis (CMV), an inflammation of the retina caused by a runaway herpes virus which, if left unchecked, invariably leads to blindness. According to Dr. Douglas Dietrich of the New York University School of Medicine, the CMV "is quite common in the population in general," estimating that 50 percent of Americans harbor the virus in their bodies. In someone with an undamaged immune system, the virus is usually harmless, but it rages out of control in a system that has been weakened or destroyed by AIDS.[16]

Ganciclovir checks the virus. It cannot destroy the CMV already present in the body, but if used indefinitely, it checks the invasion of healthy cells and preserves the eyesight of the patient. Even in the early days of the epidemic it was known that if a patient with CMV did not get ganciclovir, he went blind. If he did get the drug, he did not go blind. It was as simple as that.

What was not at all simple, however, was the slowly emerging fact that patients who used ganciclovir could not use AZT at the same time. Both ganciclovir and AZT, the first drug shown to prolong the lives of AIDS patients (see chapter 8), suppressed the ability of the bone marrow to produce blood cells, and a multiplication of the same dangerous side effect could not be tolerated by the body. Thus, some patients were faced with a harrowing choice: live longer and go blind, or live a shorter life with eyesight.

Despite this complication, the drug had a positive effect on many patients, and its value was clear. The drug was first synthesized in 1980 by Syntex, a California-based manufacturer, which in 1984 reported to the FDA that the antiviral activity of the product was highly specific to CMV. The drug looked so promising that Syntex asked the FDA for permission to distribute it free of charge until formal testing could get under way, and the FDA said yes. This was the agency at its best, exercising both common sense and compassion. Sadly, neither characteristic was much on display over the next five years as the drug traversed the torturous route of the FDA approval system.

But the FDA was not the only villain in the piece. Much of the blame for the long delay could be laid at the door of the manufacturer, itself, and the motivation, as usual, was money. As soon as the FDA said yes, Syntex began giving out the drug to any doctor who asked for it on a compassionate-use basis. At the same time, another manufacturer, Burroughs Wellcome, had developed its own version of the drug, and was also giving it away on that basis. Both companies wanted the patent on the product and the exclusive marketing rights, and between them they distributed the drug to about 6,300 people.

But neither company, at that point, was willing to invest the huge sums required for formal clinical trials of the product, not until they knew which one would get the patent. Syntex also reasoned that the FDA would not insist on the traditional placebo-controlled double-blind tests before approval. They felt that in this case such tests would be both unethical and unfeasible. Unethical because they would deprive half the subjects of a drug that was known to have value, and unfeasible because so few subjects would agree to participate, since the drug was available by other means. But the main reason was money. Syntex was simply unwilling to bet on itself until the race was over, and the result was a classic example of how not to develop a drug.

Syntex was granted the ganciclovir patent in March 1987, and promptly requested FDA approval. The answer was a flat rejection. The material that Syntex submitted to back up its application was deficient in every area required by FDA standards. The information was obscure and undetailed, dosages and regimens fluctuated wildly, and controls were virtually nonexistent. When the FDA advisory committee met to consider the Syntex application, Ellen Cooper, then the FDA scientist in charge of reviewing AIDS drugs, commented that, "What we are really left with is little better than a mass of anecdotal information."[17]

The phrase was damning, for in the world of drug approval, anecdotal information, no matter how compelling, is considered wholly unscientific and irrelevant. The phrase was also accurate. Based on traditional FDA procedure, the committee had no choice but to deny the application. As Commissioner Frank Young put it, "We can't accept testimonials."[18] Yet every member of that committee knew that ganciclovir worked, and must have known that an untold number of AIDS patients would lose their eyesight without the drug. Arno and Feiden quote one committee member as saying that the evidence in favor of ganciclovir was "close to overwhelming." Still, the FDA answer was no.

When the cold reality of the FDA decision finally sank in, everyone involved knew that a type 1 error of stunning proportion had been made. A valuable drug, effective under the circumstances and safe within the gravity of the situation, had been rejected in contravention of the FDA's stated mandate to protect the public health. The understandable reaction from the AIDS community was outrage. Even before the committee had made its final decision, a coalition of AIDS activists had demonstrated noisily in front of the FDA headquarters in Rockville, insisting on the right of a dying person to risk using any drug, approved or not, because of the possible benefits.

Inside the building, after the decision had been rendered, men and women from Syntex and the FDA met to see what could be salvaged from the situation. Looking backward was pointless. An opportunity had been lost and had to be regained. What had to be constructed was a new set of trials for ganciclovir that would supply the scientific data that had been missing from the first application. There was no further talk of using a double-blind placebo procedure. Even the FDA now agreed that, given the circumstances, such trials were unfair to those patients who received the placebo.

But the FDA was still committed to some form of placebo control, and as committees have a habit of doing, the participants proceeded to construct a camel by nailing a horse and a cow together. They named the strange creature Protocol 071, and decided that the new trials would include only patients with CMV on the periphery of the retina, where it had not yet affected the patients' sight. Everyone would get an initial intensive dose of the drug, but after that only half would receive a maintenance dose and the other half would get a placebo. The supposed safety net was that anyone on the placebo who seemed to get worse would be pulled out of the study and given the real drug.

One has to wonder how any scientist, trained in objective reasoning, could have found any essential difference between Protocol 071 and a traditional placebo-controlled trial. Patients with a deadly disease would still be denied a drug that, by common consensus, worked; and the safety net was just another way of gambling with the eyesight of the patient. A panel of ethicists came to the same conclusion—such trials would be unethical—and the plan was shelved.

What followed over the ensuing year and a half was a severe case of pharmaceutical gridlock in which all the interested parties honked their horns as loudly as possible, and nobody moved an inch. Nine months later, Syntex tried to breathe new life into

Protocol 071 by suggesting that maybe it was ethical after all. The FDA, in desperation, got onto the bandwagon, but the wagon was caught in the gridlock. A bushel of red tape descended on the project, the agency moved with a characteristic lack of dispatch, and the White House projected no sense of urgency. As time dragged on, only one of the many treatment centers meant to participate in the study was ready to enroll patients, and just about that time the FDA decided to curtail the distribution of ganciclovir under the compassionate-use program. The resultant public outcry forced the agency to reverse itself and continue distribution, but approval of the drug was still stuck in gridlock.

The eventual approval of ganciclovir was due in part to the public attention drawn to the situation by the actions of activist groups such as the AIDS Coalition to Unleash Power (ACT UP), which harassed and provoked any mainstream icon in sight. Whether it was demonstrating in front of the FDA building, raiding the headquarters of a major pharmaceutical manufacturer, invading the New York Stock Exchange, profaning the precincts of St. Patrick's Cathedral, or heckling speakers at AIDS meetings from San Francisco to Montreal, ACT UP brought the magnitude of the AIDS epidemic home to millions of Americans who would gladly have been left in ignorance. Composed primarily of gay, white males, ACT UP's posture was a combination of the shrill and the shrewd, and it brought results even when it insulted and repelled those who were working toward the same goals. ACT UP, and other advocacy groups like it, brought an outrageous spotlight to bear on an outrage, and helped to create a climate of change.

This climate of change was also brought about by the hard work of the head of NIAID, Anthony Fauci. Fauci started off as one of the favorite targets of the AIDS advocacy groups, but wound up gaining their trust and respect. He knew as well as anyone in government that glanciclovir worked; he saw the results in patients at the NIH clinics in Bethesda. He also knew that Protocol 071 was virtually dead in the water; people were simply not signing up for the trials. Fauci began to lobby everyone he could lay his hands on, including

FDA Commissioner Frank Young, pleading for a reappraisal of the drug, and as a bureaucratic insider, his pleadings carried weight.

The third reason that the drug was finally approved was that the FDA grew weary of trying to maintain a clearly untenable position. The agency had rejected a good drug on flimsy grounds, everyone knew it, and no amount of self-serving doubletalk could change that. Glanciclovir had to be approved, and if all men were angels the agency at that point would simply have thrown up its collective hands and said, "All right, we blew it. Let's fix it." But that sort of candor has always been in short supply within the bureaucracy. A charade of serious reconsideration was required, and a year and a half after the first rejection of the drug, the FDA advisory committee met again to discuss the matter.

The studies that were presented to the committee would have been thought laughable had they not already engendered so much sadness, for this was nothing more than the original data recooked and served up with a touch of finesse. Nothing had changed, no placebo-controlled trials had been performed, and much of the information was still anecdotal. But this time the committee voted unanimously to accept the application. Five years after Syntex started handing out glanciclovir on a compassionate-use basis, the drug was finally approved.

Compound Q

Compound Q is a purified protein extracted from a cucumber-like Chinese plant, and for a while it looked like one of the most promising drugs on the AIDS horizon. Used for centuries in China as a component of herbal remedies, it was first synthesized in 1987 by researchers at GeneLabs, a biotechnology firm in Redwood City, California, and was found to destroy HIV-infected cells while leaving healthy cells intact. Two years later the researchers were granted a patent on the product, known as GLQ-223, and after the preliminary animal trials had been completed the FDA gave its approval for Phase I clinical trials to begin at San Francisco General Hospital. Those trials were, of course, conducted along standard FDA guidelines, which meant that the few patients involved received only a minuscule amount of the drug as dosages were tested, checked, revised, and double-checked. The pace was slow, as it was meant to be, and what happened next could almost have been expected. Given the deep-seated distrust that the AIDS community harbored for the FDA drug-approval process, it is not surprising that a part of that community would decide to run its own clinical trials.

The organization involved in this decision was Project Inform, founded by San Francisco businessman Martin Delaney, who, along with Jim Corti and others, had organized the smuggling of unapproved pharmaceuticals from abroad during the early days of the epidemic. Ribavirin, isoprinosine, dextran sulfate—the drug of the month—whatever was deemed of value for AIDS and was unavailable because of FDA limitations was purchased in Mexico and in other countries, and brought into the United States for distribution through an underground network. Project Inform was created by Delaney to supply potential smugglers with the best routes to and from Tiajuana and Ciudad Juarez, the names of the pharmacies where the drugs could be purchased, and tips on how to get the merchandise through the U.S. Customs posts.

When the news about Compound Q surfaced in the AIDS community, Delaney, along with Larry Waites, a pediatrician, and Al Levin, an allergist, decided to run their own trials on the drug. Their reasons were twofold. They were optimistic about Q's value, and they were pessimistic about its side effects. They knew that they were dealing with a drug that could cause seizures, coma, and death, and they knew that the underground would soon be making it available illegally. Thus, given the snail's pace at which the FDA trials were proceeding, they were afraid that if the drug had any value, it would take years to get it on the market, and if the drug proved to be dangerous, too many people might be killed over those years through self-medication.

Project Inform's trials of Compound Q, which began in New York and San Francisco in May 1989, were truly underground. The study did not have the approval of the FDA; the drugs, which had been smuggled in from China, would not have met the agency's standards of purity; and, most important, the dosages used were far higher than the FDA limits imposed for Phase I testing. The underground researchers, with speed as a criterion, began with administering doses of Q that ordinarily would not have been used until Phase II of the trials, in some cases reaching levels seventeen times that of the dosage given to patients during the FDA trials at San Francisco General.

Although there were many people in the AIDS community who knew about the Project Inform trials, and the FDA, in an informal way, was aware of them, Delaney tried to maintain a veil of secrecy over the operation. He was, after all, involved in a testing scheme that was both illegal and, by the standards of the times, unethical. But any hope of preserving that secrecy disappeared in June 1989 when one of the subjects in the trials died. Robert Parr, a former Royal Air Force pilot living in San Francisco, lost consciousness two days after receiving his first dose of Q, and died in his sleep several days after that. Parr's death was never directly linked to the drug he was taking, but the obvious implications were drawn by the press, and the trials were suddenly in the public spotlight.

Delaney at once came under severe criticism from mainstream researchers, and the FDA launched an investigation into both the trials and the circumstances surrounding Parr's death. The agency position was simple and clear: the trials were illegal, and those illegal trials had probably caused the death of a subject. At any other time in the agency's history, even a year earlier, the FDA response would have been severe. The trials would have been closed down, and Project Inform would have been prosecuted to the full extent of the law. But this was 1989, and a different sort of ethos had begun to creep into the agency mindset. Instead of running out the big guns, Carl Peck, the director of the FDA's Center for Drug Evaluation, invited Delaney to a conference in which they could discuss the data gathered by Project Inform and make plans for the future.

The meeting embodied everything that Delaney reasonably could have hoped for from the FDA. Instead of being slapped in jail, he was welcomed by his onetime adversary as a colleague, more or less, and complimented on the early data that Project Inform had developed. Furthermore, he was given the green light to develop a protocol for further testing, although he was instructed to use more traditional scientific methods. Peck imposed few restrictions, the most important one being that the new trials would be conducted using GeneLabs' GLQ-223 version of the drug, and not the Chinese import. Also, Sandoz Pharmaceuticals, which holds the marketing rights to GLQ-223, contributed $250,000 to help cover the expenses of the new study.

This was a heady victory for the onetime underground pharmaceutical distributor, a legitimization of everything he had worked for. It was a historic moment for the FDA, as well, but not an easy one, for the agency had been in a no-win situation. As Arno and Feiden see it, "Had it chosen to come down hard on Project Inform, the agency would have undoubtedly been accused of being an overly rigid, heartless regulator. This time it took a more politically savvy approach and instead found itself under attack from mainstream researchers. Many were furious at the FDA's decision to have a collegial meeting with Project Inform rather than prosecuting the study organizers."[19]

An immediate critical response came from Jere Goyan, dean of the University of California at San Francisco School of Pharmacy, and himself a former FDA commissioner. "If you get these people taking these drugs willy-nilly around the country, you'll lose valuable information, and it will be at the expense of future patients."[20] A similar view was expressed by John Fletcher, who for ten years was chief of the bioethics program at NIH. Fletcher said that he was "shocked and chilled" by the Compound Q study, which, he said, "violates the very first ethical principles."[21]

A fiery critic of the Project Inform study was Arnold Relman, then editor of the *New England Journal of Medicine*, who denounced Delaney in front of five thousand people at an AIDS conference in San Francisco's Mosconi Hall. Relman criticized Delany for putting out his data before they had been examined by a peer review process, and Delaney wanted to know who had appointed Relman the guardian of what could and could not be said on scientific matters.

"This is the scientific tradition of peer review and objectivity," said Relman. Delaney replied, "How can you talk about objectivity when 89 percent of your revenues come from advertisements from drug companies?"[22]

Despite such exchanges, the trials went on, with no firm efficacy conclusions to be drawn as yet. The Project Inform data showed lower P-24 antigen levels, higher T-cells, and other improved markers, but emerging from the data was the clear indication that Compound Q had a strong positive effect on some patients, and little if any effect on others. Was it possible that the genetic makeup of each individual patient could affect the way that Compound Q was received in the body? The likelihood was there, and not just for Compound Q. In an overview of such drugs, the *Economist* pointed out that all drugs seem to work for someone. "On almost any trial, some people will get a bit better, and may become believers and proselytizers; others will feel that the drug has done nothing for them; others will die. The success of the Compound Q trial so far lies not in its results, but in the fact that it has speeded up the development of a drug that many people want to get hold of."[23] Of equal importance, the trials provided the setting for a historic move by the FDA toward a more sensitive and flexible approach to drug approval.

8

"Penicillin Couldn't Get Through That Fast"

THE FIRST DRUG APPROVED BY THE FDA SPECIFICALLY FOR USE AGAINST AIDS was azidothymidine, also known as zidovudine, and finally as AZT. In June 1985, Burroughs Wellcome, a pharmaceutical company based in Research Triangle, North Carolina, took the first formal step in gaining approval for the drug by filing an Investigational New Drug (IND) application with the agency. In March 1987, the FDA approved the drug, making it available to the public at large. The story is in the numbers. AZT went from the IND stage to full marketing approval in slightly less than two years at a time when a typical drug took eight years to go through the process. If the FDA began the 1980s committed to the traditional drug approval process, it approached the end of the decade with a flexible attitude far more in tune with the times.

But the early days of the AIDS epidemic were marked by a distinct reluctance on the part of the pharmaceutical industry to develop new drugs specifically designed to fight the disease. Given the time, effort, and cost involved in bringing a new product to market, the companies saw little reason to focus their efforts on developing drugs which, at that point, seemed destined to service a limited number of users. As we have seen, the industry prefers to present itself as an ethically motivated arm of the scientific community, reserving its persona as a commercial juggernaut for its annual reports and the privacy of the boardroom. Thus, without the incentive of a financial bonanza, the drug companies had no difficulty in ignoring a public health disaster in the making.

But that all changed when Burroughs Wellcome turned AZT, which had been developed by the government at taxpayer expense, into a cash cornucopia for its own benefit. That got the attention of the juggernaut, for AZT was the fruit of research by government scientists at the National Cancer Institute (NCI), a branch of the National Institutes for Health (NIH), but was appropriated by Burroughs under an exclusive patent for which, by 1993, it had developed a staggering $1.4 billion in revenue. Once the patent was granted, Burroughs had a lock on the drug, and soon was selling it to AIDS patients at prices ranging from $2,000 to $3,500 a year, down from an original cost of $10,000 a year because of heavy pressure from the AIDS community, and because the government-approved dosage was cut by half.[1]

But how did Burroughs Wellcome manage to get a patent on a drug that it did not discover and did not develop in the first place? The story began in 1964 when Dr. Jerome Horwitz of the Michigan Cancer Institute, operating under a grant from NCI, began screening existing drugs, looking for one that could be transformed into a cancer therapy. One of the compounds that he synthesized was AZT, the main ingredient of which was thymidine, made from the DNA in salmon and herring sperm. But AZT didn't work for cancer; it did not produce any significant antitumor activity, and so the project was shelved. Horwitz never applied for a patent, and the formula entered into the public domain. At that point, anyone who wished to could manufacture AZT.

Jump twenty years into the future. The age of AIDS had arrived, and after it had been determined that the disease was caused by a retrovirus later named HIV (human immunodeficiency virus), the NCI set up a testing center to process compounds that it hoped might work against the virus. In 1984, the institute established a Special Task Force on AIDS with Robert Gallo as scientific director

and Sam Broder as clinical director. Broder sent out a call to
pharmaceutical companies all over the country, asking them to
contribute any chemicals that might seem promising because of
their antiviral characteristics. Broder pledged to test any such
compounds at NCI, but in return he wanted a firm commitment
from the pharmaceutical companies to develop and market any
drugs that seemed to have potential. The initial response to the call
was disappointing. The drug companies simply weren't interested,
and for three reasons, one of which was financial, one of which
was emotional, and a third of which was a combination of the first
two.

On the financial end, the companies were reluctant to invest the
huge sums required to develop a drug that would be targeted at a
user community which, at that point, numbered no more than
three thousand patients. On the emotional end was fear, for there
was a very real danger involved in working with live AIDS viruses,
and the commercial scientists wanted no part of it.

A final reluctance to get involved, a combination of fear and
money, was produced by Broder's condition that the companies
agree to develop any promising drugs. Because of the risks
involved, any form of HIV research would have to take place in a
highly controlled setting. This setting was known as a P-3 level of
biosafety, and very few laboratories in the public sector, the private
sector, or in the academy qualified for that designation. The cost of
converting a laboratory to a P-3 level was close to a quarter of a
million dollars, a sizable undertaking even for the cash-rich drug
industry.[2]

But Broder was persuasive. He did not try to minimize the risk of
working with the live virus, and he did not try to minimize the cost
of converting a laboratory to a P-3 level, pointing out in passing
that his own lab in Bethesda did not meet that standard. But he
knew which button to push. He insisted over and over to drug
company executives that the AIDS epidemic was fated to spread far
beyond the minuscule patient base that then existed, that there
was no way of knowing how many people were currently infected

and how many more would be infected as time went on. He painted a picture of the future of AIDS that was distressingly pessimistic, and that turned out to be sadly accurate. It was a picture of a future green with profits, and slowly the companies began to overcome their initial reluctance to participate.

Eventually, about fifty manufacturers responded to Broder's call by sending in compounds with antiviral characteristics. Burroughs Wellcome was one of them, and one of the compounds that the company sent was Horwitz's AZT, which had been sitting on the shelf for twenty years, unpatented and in the public domain. Burroughs had tested AZT for veterinary applications, but, in the end, had chosen not to develop it. Since then it had been gathering dust until Burroughs slapped the label Compound S on it for purposes of anonymity and sent it off to NIH.

Early in 1985, after extensive testing at the NIH facilities at Bethesda, the institute discovered that AZT inhibited HIV in the test tube, the first drug ever to do so. The drug appeared to block the pathological changes that take place in infected helper T-cells, thus exposing patients to potentially fatal diseases. It was thought that further damage to the immune system could be prevented by halting viral replication, and that ways might be found to restore the immune system to health. The discovery was a tribute to the extraordinary work of the institute's Dr. Hiroaki Mitsuya, who Broder was later to call "the best viral pharmacologist in the world."[3] FDA laboratories, when requested, could not confirm the NCI findings, but an independent study group at Duke University under Dr. Dani Bolognese was able to do so.

It should be noted that at this point the scientists at Burroughs Wellcome had performed only limited scanning tests on a number of compounds, but never had worked with live HIV in the test tube or in human patients. By their own admission, they had been afraid to. It was the NCI's Broder and Mitsuya who had developed the test to determine that AZT inhibited the growth of the HIV virus, but Burroughs Wellcome did not see it that way. In June 1985, the company applied to the FDA for an Investigational New Drug (IND) for AZT, the first step in a successful campaign to take over the drug as its own. It is an indication of the desperation of the times that an FDA committee approved the application within one week, and researchers at Burroughs Wellcome, NCI, and Duke were at once able to put into motion their Phase I studies, during which a drug is tried on a limited group of people to determine its safety and its proper dosage level.

At this point, Broder had planned to send blood samples drawn from patients being medicated with AZT down to North Carolina for testing at the Burroughs Wellcome laboratories. This was a standard procedure in any cooperative effort involving the public and the private sectors, but Burroughs backed out of the arrangement only days before the tests were scheduled to begin. Burroughs admitted it was unwilling to risk handling the virus. Any sense of obligation to protect the health and safety of the American public had been overshadowed by fear.

The rapid pace continued. Phase I trials were completed within a matter of weeks, and the results were encouraging enough to allow the commencement of Phase II studies. By June 1986, almost 300 people either with AIDS or with advanced symptoms of HIV infection had been enrolled in the program. This was the traditional placebo-controlled process, but after only six months of studies it was obvious that the patients taking AZT were doing much better than those who were getting the placebo. At that point there had been 19 deaths out of the 137 patients in the placebo group, with only 1 death out of 145 in the AZT group. The difference was so dramatic that in September 1986, tradition was tossed aside and all patients in the placebo group were offered the actual drug; and to keep up the pace, Burroughs was told that Phase III studies would not be required.

In retrospect, one would imagine that the rapid progress of AZT through the approval process would have brought cheers of encouragement from AIDS advocacy groups, but drugs such as aerosolized pentamidine and ganciclovir (see chapter 7) were still caught in a pipeline that seemed clogged with endless bureaucratic

delays, and to some minds the FDA could do no good. According to
Shilts, "There was talk that no treatment or vaccine would get
quick FDA approval for experimentation unless it was developed
by the federal government; among AIDS organizers, NIH had
become the acronym for the agency disinterested in treatments
that were Not Invented Here. There was no stop the government
did not pull out for AZT, a drug the NCI had originally developed.
However, there seemed no bit of red tape too minor to delay the
release of other treatments."[4]

As an argument, this was understandable, but of questionable
merit. The FDA was making all of the right moves with AZT, and all
the wrong ones with the other drugs. It would take the most
simplistic of interpretations to build a conspiracy out of a situation
so commonplace in the bureaucracy, but the moment had need of
a devil, and the FDA, on past performance, qualified for the job.

The AZT program went on apace. With the Phase III studies
waived, all that was needed was a New Drug Application (NDA), a
document so lengthy and complex that a single application has
been known to fill an entire trailer truck. Burroughs submitted the
application to the FDA in stages, with the final installment
delivered on December 2, 1986, and six weeks later the FDA
advisory committee met to consider the merits of what had been
presented.

There were many sticking points that argued against approval,
all derived from the fact that the drug had been tested during just
one major trial that had lasted only six months. Because of that, no
one really knew very much about AZT. No one knew how long an
effective life the drug would have, what its toxic effects might
prove to be like in the future, or even what the ideal dosage would
be. (The arbitrary dosage decided on was, in fact, later lowered by
50 percent.) Also, no one could say with any certainty just who
would benefit from the drug.

Much was made of the fact that of the roughly three hundred
people involved in the trials, only thirteen had been women. But
the number of new cases of AIDS in women of reproductive age

was increasing at an alarming rate, and AIDS was fast becoming the leading cause of death for women between the ages of twenty and forty in the major cities of North and South America, Western Europe, and sub-Saharan Africa. In the United States, AIDS had hit hardest among black and Hispanic women, and while these women represented only 17 percent of the female population, they made up 73 percent of women with AIDS. AIDS was also having a devastating impact on infant mortality, since over 80 percent of HIV-infected children under the age of thirteen had acquired HIV from their infected mothers, and it was estimated that between 24 and 33 percent of children born to infected women would develop the disease.[5]

Despite this, the patient population of the trials had been composed almost exclusively of one social and racial group, gay white men, and because of that, legitimate questions were raised about the validity of the data collected. In light of this, should the drug be approved for use only by the limited population on which it had been tested? The answer was no for two reasons. Every member of the committee knew that once a drug is licensed, physicians legally can, and often will, do exactly what they wish with it. Once licensed, the drug could not be kept from the general population. Every member of the committee also knew that approval of the drug for a narrow population range would be an act so politically explosive that it could not be considered. The FDA had already weathered the slings and arrows of ACT UP demonstrations on the streets of Rockville, and it was not about to expose itself to similar actions from militant groups of women and minorities.

In the end, the reservations held by many of the committee members melted away, and the drug was approved by a vote of ten-to-one. Many members of the panel, aware of the need for approval, still voted with their fingers crossed. And they were not alone. The traditional principles of drug approval had been distorted beyond recognition, and James Todd, senior vice-president of the American Medical Association, reacted to the vote by noting that, "Penicillin couldn't get through that fast."[6]

But for all the carping, the approval of AZT was an innovative action by the FDA that was responsive to the needs of the times. Instead of sitting back and waiting for things to happen, the agency had gone out and helped them to happen by following the progress of the Phase I and Phase II trials, and reviewing the data at every step along the way. No matter what one might think about the decision to approve, no one could say that the agency had been in any way dilatory, ill-informed, or hidebound.

The atmosphere created by the rapid approval of AZT in March 1987 was pervasive, and led to the creation in the same month of a new category of experimental drugs known as treatment investigational new drugs (treatment INDs). The category was limited to drugs designed for life-threatening diseases, and the maker had to meet three criteria:

1) That no other treatment was available.
2) That the drug was not ineffective.
3) That the drug was not unreasonably dangerous.

If those criteria could be met, and if the manufacturer guaranteed to continue with ongoing research, then the FDA would permit the drug to be sold after Phase I trials had been completed.

On the surface, this would seem like a quantum leap away from established agency procedure. To require that a drug be "not ineffective" was far removed from demonstrating efficacy, and to require that it not be "unreasonably dangerous" was light-years away from demonstrating safety. Not surprisingly, the initial reaction throughout most of the AIDS advocacy community was positive and enthusiastic. Conversely, the medical community expressed its displeasure with the concept, citing the dangers involved, and warning the public against quacks and frauds. On Capitol Hill, the reaction demonstrated a dichotomy of feeling among those who cared deeply about people with AIDS (by no means a majority of either the House or the Senate) and who, at the same time, cared deeply about consumer protection, which meant keeping unsafe drugs off the market.

Lost, for the moment, in the conflicting reactions to the announcement was the fact that there was nothing terribly new about treatment INDs. Over the years the FDA had routinely granted permission to large groups of patients to use experimental drugs, and that permission had not been limited to people who were dying. In fact, thalidomide, the drug that caused so many birth defects abroad and led to the passage of the Kefauver Amendments, had been prescribed for almost twenty thousand patients in the United States under such a program. In the same sense, the FDA in the 1960s had given conditional approval for the drug L-dopa (for Parkinson's disease), and in the 1970s had approved methadone in the same fashion.

Thus, by creating treatment INDs for AIDS and other fatal diseases, the FDA was actually adopting a more restricted version of a practice that had been in place for decades. The result was more a shift of emphasis than anything else, and even as such it was a worthwhile move, but it did not break new ground. Peter Barton Hutt, a Washington lawyer and former chief counsel at the FDA, concluded that, "There has not been a single new idea in the regulation of new drugs since the 1970s," and Martin Delaney, once an enthusiast for treatment INDs, said that, "I don't think the progress that we hoped for has been made. People still can't get the drugs they need."[7]

Two years later came the concept of "parallel track" testing, the brainchild of Anthony Fauci of NIAID. The category was designed to meet the needs of AIDS patients who could not join a formal drug study program, and thus receive the drug involved, for a variety of reasons. Some could not get in because a particular study was fully enrolled, some because they could not meet the eligibility criteria, and some because they lived too far away from the study center. Under the parallel track program, these patients would receive the drug from their personal physicians, and their reactions would be observed separately from those in the more closely controlled clinical trials.

Once again, the new program did not break any new ground, but provided a shift in emphasis that was welcomed by AIDS advocacy groups. The FDA itself was something less than fully enthusiastic about the idea since it had originated at NIAID, a bit of political poaching that was not quite cricket even by Washington standards. The agency felt, and made public, the feeling that parallel track was redundant, and that treatment INDs covered the same goals and needs. Still, the FDA approved the program, and in October 1992 it announced that the first AIDS drug, d4T (stavudine), had been made available under parallel track.[8]

While the relatively rapid FDA approval of AZT was hailed as the dawn of a new era, the working partnership between NCI and Burroughs Wellcome was also hailed at the time as a model of how the private and the public sectors could cooperate in times of crisis. This, however, was a short-lived honeymoon. Despite the fact that government scientists, Broder and Mitsuya, had been responsible for the development of AZT, Burroughs adopted the position that the basic concept of using the drug as a treatment for HIV infection had been made by its own people, and the company applied for an exclusive patent on the product. It was a bold move that was several times rejected by the U.S. Patent Office, and eventually approved for reasons that have never been made clear. Arno and Freiden suggest some form of chicanery, claiming that, "It is known that key documents likely to explain the shift in thinking at the patent office are missing, and there are hints that political pressures were brought to bear."[9] Considering the climate engendered during the Reagan-Bush years, it is not an unlikely possibility.

AZT, marketed as Retrovir by Burroughs Wellcome, thus became the first antiretroviral drug indicated for the treatment of adults and children with HIV infection. Burroughs began marketing Retrovir in the United States in March 1987, and by 1993 the federal government had paid the manufacturer about $500 million in fees without receiving any royalties in return.

But Burroughs Wellcome was not the only pharmaceutical

company that was enjoying a free ride at taxpayer expense. According to a report by the Center for the Study of Responsive Law, the majority of what the FDA deemed the most important and innovative drugs coming to the market in 1990 and 1991 were developed with substantial help from NIH. The study also found that thirty-four of thirty-seven cancer drugs that came on the market since 1955, when NCI began its search for such drugs, were developed with significant federal financing.[10]

One example was Ceredase, a drug used to treat Gaucher's disease, which cost $300,000 a year. The drug was discovered and developed by federal scientists and academic researchers supported by federal grants, and then turned over to a small Boston company, Genzyme, for marketing. Another drug, Levamisole, was marketed for years to cure sheep of worm infestations, but then turned out to be remarkably successful in treating colon cancer in humans. The studies that determined its effectiveness against cancer were conducted entirely by government scientists, but Johnson & Johnson, which marketed both versions of the drug, charged one hundred times as much for the human version as it did for the identical drug for sheep. And then, as if to show its total contempt for the American taxpayer, the company turned around and marketed the drug in Europe at a price considerably lower than that which it charged in the United States. Johnson & Johnson reacted to criticism by claiming that the cost of training doctors to use Levamisole was the reason for the higher price for the human version, but critics dismissed the "training" costs as advertising and promotion.

As of 1989, as a direct result of the AZT controversy, NIH began to insert a clause in its contracts with pharmaceutical companies that required the companies to charge a "fair" price for the drugs that they had developed in conjunction with the government. But NIH soon admitted that it had no way of determining whether or not a price was fair, and no way to enforce such a price even if it knew what one was. This was confirmed by a senior analyst at the Office of Technology Assessment, who concluded that NIH had "neither the legal authority nor any true procedure, or perhaps even the incentive to implement that policy."[11]

When this situation became clear, it was suggested that instead of handing out licenses to market the drugs, NIH should hold an auction or, failing that, charge the drug companies higher royalties. The government can ask for royalties whenever it licenses a drug to a pharmaceutical company, but traditionally it has asked for, and received, very little. From 1981 to 1990, the government collected $26 million in royalties, of which $22.6 million came from only two licenses. These were for the test for detecting antibodies to the virus that causes AIDS and the hepatitis B vaccine. In 1990, the health institutes collected $5.84 million from their patents, which was less than four-tenths of 1 percent of the amount they spent on research and development by their own scientists.[12]

Eventually it was a generic drug company, the much harried Barr Laboratories, that decided to challenge the exclusivity of the Burroughs Wellcome patent for AZT. Barr got into the game after a consumer advocacy organization, Public Citizen Litigation Group, filed suit against Burroughs on behalf of an AIDS health group and two HIV patients. The purpose was to invalidate the Burroughs patent for AZT, which would allow other firms to make and market the drug, a sure way to lower the price. The Public Citizen suit named the U.S. government as one of the defendants because of its ownership interests in the patent, hoping that the government would take a position in the case, and that hope was quickly realized.

Dr. Bernadine Healy, then the director of NIH, issued a statement agreeing with the position of Public Citizen, and claiming for Drs. Broder and Mitsuya the right to have their names listed as co-inventors of AZT. The Public Citizen suit was dismissed by the courts on the grounds that it and the co-plaintiffs lacked the legal standing to sue, and it was at that point that Barr entered the battle. Backed by the AIDS community, which foresaw a dramatic drop in the price of the drug if a generic version could be marketed, Barr obtained a nonexclusive license from NIH to do just that. Barr agreed to sell its product at approximately 50 percent of the Burroughs price, unless that figure failed to cover its costs. Barr, in fact, offered to specify an actual price in its agreement with NIH, but the government declined the offer, apparently on ideological grounds.

Once Barr obtained the license from NIH, it notified Burroughs, as required by law, that it intended to manufacture a generic version of AZT, and Burroughs then brought suit against Barr to prevent the infringement of its patent. An additional complication developed when the NIH apparently had second thoughts about using Barr as its surrogate. The firm, after all, had had more than its share of problems with the FDA, and who was to say if those problems might not continue? The NIH decided to hedge its bet by making a similar deal with another generic manufacturer, Novopharm Inc., thus precipitating another court battle. There was, however, a subtle distinction between the two lawsuits in that while both generic firms were trying to have the Burroughs patent declared invalid, Barr was looking only to share the rights with Burroughs, while Novopharm wanted them all to itself.

But in July 1993, a federal judge cut short the legal battle when he upheld Burroughs Wellcome's patent protection for AZT. U.S. District Judge Malcom Howard issued a directed verdict in favor of the company, declaring Burroughs Wellcome scientists the sole inventors of the treatment. In delivering the verdict, the judge said that he could no longer find a legal basis for the case against Burroughs Wellcome. Attorneys for Barr and Novopharm said that they would appeal the case.[13]

But this news, bad as it was for AIDS activists, was overshadowed at the time by a preliminary report of a large European drug trial that raised perplexing new questions about AZT. The report was promptly dubbed the "Concorde" because, like the airplane of the same name, it was largely an Anglo-French collaboration, and it suggested that people in the early stage of HIV infection got little, if any, benefit from the use of AZT. These findings ran counter to the belief among most American AIDS experts that AZT delayed the onset of AIDS and should therefore be prescribed before symptoms develop and when the only evidence of infection is blood-test abnormality.

The report indicated that treating asymptomatic HIV patients with AZT did not delay either AIDS or death during a three-year period of observation. "Persons taking the drug showed a slightly lower risk of developing AIDS or dying within one year of starting treatment, but that advantage was gone two years later. At the same time, persons on AZT had a higher risk of side effects—such as anemia and nausea—than persons not taking the drug."[14]

In the Concorde trial, 1,749 people were randomly assigned to take either AZT or a placebo until they developed any symptoms of infection. Whenever that occurred, patients taking the placebo were switched to AZT. After three years, 18 percent of each group had progressed to AIDS, and during that same time, 8 percent of the AZT group and 7 percent of the placebo group had died. Taking AZT, apparently, had little effect at that stage of the disease.[15]

Ian Weller, a professor at University College, London, Medical School, and the head of the English part of the study, anticipated the American reaction to the report. "I think it's going to be a very unpalatable result, and perhaps an unpopular one. I think some people are not going to believe it."[16]

The Concorde study did find that the CD4 cell count in the group taking AZT was slightly higher after one year, and slightly fewer people in that group had progressed to AIDS. This was explained as the result of a small but immediate rise in CD4 cells—a subclass of infection-fighting white blood cells—when AZT is begun. But Weller discounted that piece of evidence, arguing that too much significance had been given to CD4 counts as measures of health in previous HIV studies. "These biological changes that you see do not have a clinical payoff," he said.[17]

Weller was correct in predicting the American reaction. U.S. AIDS researchers at once predicted that the Concorde study would have little impact. Because it was sufficiently different from

American trials, they believed the study might not be viewed as comparable; and the head of NIAID, Daniel Hoth, cautioned against using the preliminary report to change the current medical practice. But the report was not received in the United States with total surprise. Many of those who had favored approval of AZT with their fingers crossed had expected the arrival of something like the Concorde someday. The original FDA-sponsored trial of six months, and three subsequent confirming trials, had simply been too short. The Concorde report, in fact, contained more "end points"—death or progression of the illness from early infection to AIDS—than all four American trials combined.[18]

But no one was voicing any regrets over the approval of AZT in 1987. Even if the value of the drug in the early stages of infection had now been questioned, AZT was still, in many cases, a working and valuable therapy. Its quick approval had been an imperative action demanded by the times, and that approval had set the stage for a further easing of the FDA approval process. It had also prepared the agency for the next major AIDS drug to come down the pike. What the FDA learned from the AZT experience in 1987 would serve the agency well in 1991 when Bristol-Myers Squibb presented the drug dideoxyinosine (ddI) for approval.

9

Thicker Than Water

POPE JOHN PAUL II, ON A TRIP TO THE UNITED STATES, WRAPPED HIS ARMS around seven-year-old Brendan O'Rourke in an embrace of compassion. The boy had AIDS, received from a blood transfusion.

Belinda Mason, a self-described "hillbilly from eastern Kentucky," was appointed to the National Commission on AIDS. In the days before Magic Johnson, she was the only member of the commission who actually had the disease, and she had it because of a blood transfusion.

The pregnant wife of former "Starsky & Hutch" star, Paul Michael Glaser, was infected with AIDS from a contaminated blood transfusion during delivery, and unwittingly passed it on to her daughter during breast-feeding.

Tennis legend Arthur Ashe died of AIDS at the age of forty-nine, ten years after receiving a contaminated transfusion.

These were some of the stories about blood transfusions that made the headlines in the 1980s and into the 1990s, and these were the images that generated fear. During the decade an average of 4 million people in the United States received blood transfusions each year, and some of the blood they received was tainted. As of September 1991, the Centers for Disease Control in Atlanta had reported a cumulative total of 4,428 cases of transfusion AIDS, in addition to 4,412 cases in which transfusion was one of the risks to which the victim had been exposed.[1] The CDC also estimated that between 70 and 90 percent of the nation's hemophiliacs had been infected by contaminated blood. Many of those people had died, many others were doomed, and no one could say with any certainty how many of those deaths could have been avoided. But the story of blood transfusions in the eighties was the story of how greed and sloth on the part of the blood banking industry wasted three precious years and created an epidemic within an epidemic. It was also the story of how the FDA allowed it to happen.

The responsibility of the FDA in terms of blood transfusions was clear; the agency was mandated to protect the safety of the national blood supply. In the 1980s, it exercised that mandate through the use of the Blood Products Advisory Committee, a small group of industry and medical experts that met quarterly with FDA officials to offer advice on regulatory matters. As had happened so often in the past, the agency was relying on the opinions of the industry it was supposed to be regulating. In early December 1982, an unscheduled meeting of the committee was hastily called at the National Library of Medicine in Bethesda, Maryland, and over a sandwich lunch the members listened to a presentation by Dr. Bruce Evatt of the Centers for Disease Control, the Atlanta-based federal agency that tracked clues about the spread of infectious diseases.

In this case, the disease in question was acquired immune deficiency syndrome, known even in those early days by the acronym AIDS. The immunological disorder had been a CDC concern since 1981, and Evatt began by offering data from CDC investigations of several unusual cases of AIDS. In each of them, the victims were not homosexual men, intravenous drug users, or members of any other group known at the time to be at high risk for the disease. But they did have one thing in common: they all either had received blood transfusions or had used a blood-based product within the previous five years. The concept of a coincidence had been too much to stomach, and CDC investigators had traced back through blood bank records to identify the donors of blood transfused to the AIDS victims. Again, in each case a common factor had been discovered: at least one donor in each case had been in an AIDS high-risk group or had developed the disease. Clearly, whatever was causing AIDS was an infectious agent, probably a virus, and it had begun to contaminate the nation's blood supply.[2]

One month later the National Hemophilia Foundation issued a public statement asking blood and plasma collectors to screen all donors and to discourage donations from AIDS high-risk groups. Since hemophiliacs are the most vulnerable recipients of donated blood, the statement set off alarm bells within that community.

Each year the U.S. blood banking industry collects about 14 million units of blood, which are then broken down into 42 million units of red cells, platelets, and plasma to be sold on the open market at competitive prices. The organizations that collect the blood fall into two categories. One is composed of three nonprofit organizations, the Red Cross, the American Association of Blood Banks (AABB), and the Council of Community Blood Centers (CCBC), all of which collect whole blood from unpaid donors. The other group is made up of commercial companies that collect only plasma from paid donors, returning the red blood cells to the donor during the procedure. The plasma is then broken down into products such as gamma globulin, and clotting factor VIII for hemophiliacs. This side of the business is dominated by pharmaceutical giants such as Alpha, Armour, Cutter, and Hyland.

The Red Cross and most of the other whole blood collectors are known as "not-for-profit" organizations, but this only means that they have met the requirements of the Internal Revenue Service that qualify them for tax-exempt status. They are, in fact, highly profitable. For the year ending June 1990, the Blood Service Division of the American Red Cross posted revenues of just under $700 million, a figure that, were it a for-profit corporation, would place it near the outside edge of the Fortune

500 list.[3] If the Red Cross were publicly owned, you would want to have stock in it.

The behavior of the two different blood industries during the early days of the AIDS crisis, as pointed out by Joseph Feldschuh in his book *Safe Blood*, exposed a fundamental flaw in the system. While neither of the groups acted in a particularly heroic manner, it was the voluntary, tax-exempt organizations—the Red Cross, the AABB, and the CCBC—that clearly evaded their responsibility to the public.[4] And in those early days the FDA was equally guilty.

The explosive news presented by the CDC in December 1982 should have shocked the blood bankers and the FDA into immediate action. The public should have been told in the strongest possible terms about the contamination of the blood supply, and immediate steps should have been taken to protect the safety of that supply.

Instead, the FDA took no immediate action to enforce any donor guidelines on the blood industry. According to Randy Shilts in *And The Band Played On*, some FDA regulators resented the CDC's brash invasion of what was plainly their territory, the blood industry.

> Moreover, many at the FDA did not believe that this so-called epidemic of immune supression even existed. Privately, in conversations with CDC officials, FDA officials confided that they thought the CDC had taken a bunch of unrelated illnesses and lumped them into some made-up phenomenon as a brazen ruse to get publicity and funding for their threatened agency. Bureaucrats have been known to undertake more questionable methods to protect their budgets. Given the Reagan administration's wholesale budget slashing, this would not be all that drastic a reaction.[5]

In the end, everybody agreed that they should do one thing: Wait and see what happens. The situation would clarify itself and then they would move. How could the government be expected to forge national policy for more than 220 million Americans just because three hemophiliacs got sick?[6]

But while the FDA was waiting to see what would happen, the Red Cross and the other tax-exempt bankers were busily protecting their interests in a totally shameless fashion. Despite the hard evidence displayed by the CDC, the bankers at first simply refused to admit that AIDS could be transmitted by transfusion. Then, when they were finally forced to allow the possibility, they downplayed the risks involved. They delayed putting in place an active screening procedure that would discourage blood donations from people in AIDS high-risk groups, they resisted the use of surrogate blood tests, and they made exaggerated claims about the ability of the tests then in place to identify AIDS-contaminated blood.

The plasma sector, on the other hand, immediately agreed to comply with the request of the National Hemophilia Foundation. According to Feldschuh, "Some commercial blood plasma purchasers did not even wait for the National Hemophilia Foundation to request the more exacting screening procedures. As early as December 1982, Armour began buying blood from low-risk suppliers and stopped buying it from high-risk areas like New York, San Francisco, and Los Angeles."[7] At the same time the entire commercial plasma sector put into place new procedures for screening donors.

"Meanwhile, the Red Cross and the other tax-exempt organizations refused to accept publicly the notion that AIDS could be transmitted by transfusion. They argued that the evidence remained 'incomplete' and 'inconclusive.'"[8] Over a year after the original CDC warning, Aaron Kellner, president of the New York Blood Center, was still insisting that, "We're not convinced that AIDS is transmitted by blood transfusion. . . . The evidence is still very shaky."[9]

It would be pleasant to believe that this reluctance to face facts was nothing more than scientific conservatism, but one is forced to the conclusion that the motivation was unconnected with science. It was done, quite simply, for money. Blood is big business. Blood bankers sell the blood they collect from the public on the open market to hospitals and patients. Even in the early eighties, it was clearly known that there was no danger of a donor contracting AIDS during the act of giving, but the blood bankers were concerned to the point of obsession that no stigma of any kind be attached to any part of the transfusion process. They were convinced that if, somehow, the public got the idea that giving blood was risky, donations would drop dramatically. Blood was their product, and if there was no product, there was no profit.

If the compelling motive of the tax-free blood bankers was greed, the FDA position in these early days was fueled only by a reluctance to offend the industry it was supposed to be regulating. Later on, under the pressure of public opinion, and despite the Reagan-Bush administrations' reluctance to involve themselves more than minimally in anything concerning AIDS, the FDA would perform a creditable job. But in those early days the agency was more concerned with not rocking the boat than with saving victims.

In March 1983, the FDA did issue screening recommendations to the Red Cross and other tax-exempt organizations, but these suggestions were of the simplest sort, and, besides, most blood bankers simply ignored them. The FDA recommendations did not include a physical examination, they did not include direct questioning of donors, and they did not include the signing of an affidavit. In the meantime, says Feldschuh, "1.5 million units of blood had been collected between January and March 1983, and no one will ever know how much contaminated blood slipped through because even these simple guidelines were not utilized."[10]

It was slowly becoming clear to some of the more advanced thinkers in the blood business that screening out donors from high-risk groups would not be enough. The actual blood would have to be screened, a difficult task because at that point the actual AIDS virus had not been identified. Instead, it would be necessary to find a surrogate test that would detect a white cell abnormality that was often found in the blood of AIDS patients. This was the T-cell test. If a donor's blood showed this T-cell abnormality, one could assume that it was infected with AIDS, and discard it.

It would seem, with the benefit of hindsight, that such an eminently logical and simple test would have been at once embraced by the blood industry, and at once insisted upon by the FDA. Instead, the industry mounted a stiff opposition to any such testing, and when it came to the crunch the FDA folded. Dr. Edgar Engleman, a Stanford blood banker, was one of the first to advocate surrogate testing, and in the summer of 1983 he submitted such a recommendation to the annual meeting of the AABB. The recommendation was rejected out of hand, and years later he recalled the reaction that he received: "It was particularly distressing to discover . . . that the subject of transfusion-associated AIDS wasn't even on the meeting's program. While we certainly don't believe that there was a conscious conspiracy to repress information . . . there seemed to be great reluctance to acknowledge the problem."[11]

Finally, in December 1983 the FDA shook itself from its lethargy and recommended to the blood industry a somewhat different surrogate test involving core antibodies, but the Red Cross and the AABB once again mounted stiff resistance, maintaining that nationwide surrogate testing was unnecessary, and once again the FDA avoided making a decision by appointing a panel to study the matter further. More millions of units of blood were collected, much of it unscreened, and transfused into patients, until March 1985, when the FDA approved and enforced the first AIDS antibody screen test, and nationwide testing began. Three years had gone by since the medical community, including the FDA, had been told that the HIV virus could be communicated by transfusion—three wasted years and nobody knows how many wasted lives.

March 1985 marked the turning point for the FDA in the blood-AIDS crisis. With nationwide testing in place, the agency was able to turn its attention to the supervision and regulation of blood bank centers, a job requiring little imagination but a great deal of diligence. In this, the agency proved to be singularly successful, but often found itself pitted against its onetime ally, the American Red Cross.

The Red Cross in America is as close as we get to having a sacred cow grazing on the lawn. Just mention the name and images flash of a soup kitchen in a flooded Texas town, Bangladesh after a cyclone, packages for prisoners of war, refugee camps in Jordan. The Red Cross is nickels and dimes at collection time, fund-raisers at fun fairs, blood drives and give-till-it-hurts. The Red Cross is apple pie and apron strings, the Fourth of July and Thanksgiving rolled into one. To many Americans it has represented all that is warm and generous in the national spirit. An aura hangs over the organization, which makes it difficult to criticize. But then there were all those headlines.

RED CROSS CLOSING D.C. BLOOD BANK. "In February 1988, 24 units of contaminated or questionable blood from the E Street laboratory passed through safety checks and were sent out for transfusion."[12]

BLOOD BANK IN N.Y. WILL LOSE LICENSE. "The FDA informed Red Cross management that it was taking the action because the center had repeatedly failed to ensure a safe blood supply."[13]

BLOOD BANK FACES OREGON SHUTDOWN. "The FDA said the center had repeatedly accepted donations from people ineligible to donate, like drug users."[14]

RED CROSS FAULTED ON TAINTED-BLOOD REPORTS. "The American Red Cross has delayed reporting or has failed to report to the federal government hundreds of errors it has made in handling contaminated blood."[15]

In fact, the American Red Cross, which collects more than half of the whole blood donated in the United States, had serious and persistent problems during the 1980s with its procedures for testing and keeping track of blood. According to FDA inspection reports, during that period various Red Cross centers released infected blood into the national supply, violated AIDS testing procedures, and failed to deter infected or high-risk donors such as homosexuals and IV users. In fact, an internal Red Cross document revealed that in just one six-month period in 1988, more than 2,400 blood products were released to the public that had not been properly screened for AIDS or hepatitis.[16]

And the Red Cross was not the only blood collection agency
with problems. The condition was pervasive, with the FDA citing
the New York Blood Center, an independent bank, for numerous
faults in its computer system and for mislabeling shipments of
blood. The government report was peppered with examples of
blood shipments to hospitals that were returned to the center
because of incorrect or missing labels, and closed with the
damning comment that the "labels are not examined for accuracy
upon receipt and before release to inventory."[17]

As the 1980s came to a close it was clear that a risk-free system
of blood collection, if such a system could be said to exist, was still
far in the future. The odds on contracting AIDS seemed
unacceptably high to many people, and opinions varied as to what
those odds actually were.

The American Red Cross, with its Pollyannish history of
optimism on the subject, quoted the beguiling odds of one in
153,000, but this assessment was suspect, to say the least, because
the Red Cross did not collect blood in New York City, San
Francisco, and large parts of Texas, thus omitting some of the
highest-risk AIDS areas from their findings.[18]

A more realistic figure of one in 40,000 was quoted by the CDC,
but, in point of fact, nobody really knew. The question defied a
simple answer because of the long incubation period of the HIV
virus, which typically takes about seven years to destroy the
immune system and present itself as a clinical case of AIDS. As a
result, even the most serious of studies had to rely on computer-
generated estimates, and while the CDC figure of one in 40,000
was widely accepted, that number stated the risk of infection in
terms of a single unit of blood. Since the average transfusion
requires 5.4 units, and since those units come from different
donors, the risk to the patient was closer to one in 8,000.[19] Not bad
odds for a lottery ticket, but would you bet your life on it?

Late in the decade, some people began to say no to the odds,
and one of the more knowing comments heard when Pope John
Paul II embraced young Brendan O'Rourke was that His Holiness

did not have to worry about transfusion AIDS. The pope, like a growing number of statesmen, dignitaries, and celebrities, carried a supply of his own blood with him, drawn from his veins and frozen for future use in an emergency. Called autologous donation, it was, and still is, the safest form of transfusion. Aside from the pope, presidents and rock stars soon were doing it, and autologous donation became the latest mark of medical sophistication. ("You mean you aren't storing your own blood yet? I thought that everyone was doing it.")

"Modern techniques of freezing allow for the safe storage of blood for up to twenty years, although the FDA limits us to ten," said Feldschuh, the director of Idant, the first public autologous blood program in the country. "You can't contract AIDS or hepatitis or suffer any kind of transfusion reaction from your own blood."[20]

Idant and other such centers also stored blood for designated donation, a procedure in which family members or close friends with the appropriate blood types contributed units for a particular person about to undergo surgery. In a more sophisticated version of the plan, the donors would "network" the blood for the use of any member who might need it in the future. On the surface, the idea sounded eminently sensible. After all, what blood would you trust more than that which comes from the veins of your nearest and dearest? But the blood banking industry, the FDA, and some physicians did not see it that way, and they based their opposition on grounds that they claimed were both moral and medical.

Morally, their argument ran that donated blood was the property of the community, and that to insist on a specific recipient was to commit an act of social selfishness. (Translation: Blood banks sell the blood they collect at an average price of $60 per unit, and every unit that is not sold through a bank means money out of the banker's pocket.)

The medical argument, as presented by Dr. Jay Epstein, acting deputy director of the FDA's Division of Transfusion Science, was that, "The ability to ascertain risk in a family member is no better than the ability to ascertain risk in a nonfamily member. The family member may have psychological reasons for denying the risk involved."[21] (Translation: For all you know, your cousin Eddie may be a closet homosexual, and your Aunt Irene may be shooting up dope.)

But, as if recognizing the weakness of the position, Epstein added, "Most directors of medical centers would probably be sympathetic to the use of a close family member, but the further you get from the center of the family, the less tenable is the argument."[22]

Which brings us back to young Brendan O'Rourke, standing with the pope's arms wrapped around him. As an infant, Brendan needed surgery, and his parents wanted to provide their own blood. What parents wouldn't? But the physician and the hospital refused to allow it. It was against hospital policy. It was socially selfish. Their blood belonged to the community, not to their son.

Brendan's parents fought in the courts, and Brendan's parents lost. Brendan received blood from eighteen different donors, one of whom gave him AIDS. Seven years later he was dead.

Only one in 153,000? Only one in 40,000? Only one in 8,000? No one really knew, but even the FDA, which, in terms of blood safety, could be as sanguine as the Red Cross, conceded the risk factor in blood transfusions, but insisted that those risks were low when compared to certain other medical interventions such as taking penicillin or using anesthesia. "You have to look at blood as you would any other medical therapeutic," said Epstein. "Blood is a drug, and it carries certain risks."[23]

In order to minimize those risks, by the end of the decade all blood banks were screening potential donors before they were allowed to give blood. "Just ten years ago, giving blood was fairly straightforward," recalled Dr. Gerald Sandler, then medical director at the American Red Cross Blood Services. "Donors had to answer a few questions about hepatitis, malaria, and overseas travel. If people felt well, had normal blood pressure, and were not anemic, they were likely to finish donating blood and be munching donuts in twenty minutes."[24]

The process of checking the safety of donated blood is far more complex today. Every potential donor is asked to read a brochure about the risks of infection by blood-borne viruses, and then must answer specific and graphic questions about his or her health, use of drugs with intravenous needles, and sexual behavior. (Are you a man who has had sex with another man since 1977, even one time? Are you a man who has had sex with a female prostitute or a woman who has had sex with a male prostitute in the past six months?) If the answers are satisfactory, the blood is accepted for donation and testing.

But satisfactory answers are no guarantee that the prospective donor is safe, and a puzzling question is why members of high-risk groups continue to attempt to donate blood. It isn't for money—whole-blood donors aren't paid—and it would be a rare such person who did not know the risks involved in giving his or her blood. Still, they continue to do it, and one explanation of why they do it is the social pressure that accompanies the periodic blood drives in communities all over the country.

"The pressure to give is severe," said one would-be donor who insisted on anonymity. "No one in the office knew that I was gay, and I wanted to keep it that way. I mean, I had to keep it that way, and if I had refused to give blood I would have stuck out like a sore thumb. Everyone was doing it."[25]

What happened?

"I lied on the questionnaire and I lied at the interview. I gave my blood, it was screened, and I was rejected."[26]

Had you known that you were positive?

"No, that's how I found out. That's another reason why some high-risk people try to donate. It's a free and confidential way of finding out if you're positive."[27]

What would happen if you tried to donate again?

"They'd turn me down. I'm listed in the Donor Deferral registry. But nobody knows that. It's confidential."[28]

And what's going to happen at the next blood drive?

He shrugged.

All blood banks today test for the presence of HIV antibodies, an indication of AIDS, and they also test for three forms of hepatitis, for HTLV-1, a virus that causes a rare form of leukemia, and for the liver enzyme ALT. But it is the AIDS testing that occupies the center of the public consciousness, primarily because the test in use at blood centers does not detect the AIDS virus itself, but only the AIDS antibodies produced by the virus. Since there is a significant delay between infection with the AIDS virus and the appearance of AIDS antibodies—from as little as eight weeks to more than a year— an unknown number of AIDS-infected units of blood regularly slip into the public supply during this "window" period. These are units that produce the statistical estimates, but even if an absolutely foolproof test for the AIDS virus itself were to appear tomorrow, the blood supply would still be at risk because of the large number of patients who receive blood contaminated by the hepatitis viruses.

In testimony before Congress, Dr. Ross D. Eckert of Claremont McKenna College stated that, "For the past several years, about 500 patients a day were infected with hepatitis viruses, and about 4,000 of them per year will develop fatal cirrhosis within five to ten years. Losing 4,000 people a year is like losing a fully loaded DC-10 each month."[29]

But for all the legitimate concern about the dangers of hepatitis it is the fear of AIDS that brings out the monsters that stalk in the night. You're just as dead from cirrhosis of the liver, but AIDS provokes a fear like no other. AIDS is more than a death-dealing disease. AIDS is a stigma so closely associated in the public mind with homosexuality and IV drug use that then-President Bush felt called upon to describe it as "a disease where you can control its spread by your own personal behavior."[30] Not very comforting words to the tens of thousands who had contracted AIDS from blood transfusion.

"Tell me, Mr. President, where did I go wrong?" reflected one of those thousands. "How could I have changed my personal behavior? All I was trying to do was to stay alive."[31]

After almost a decade of greed, sloth, and mismanagement of the nation's blood supply, the House subcommittee chaired by Representative John Dingell convened on July 13, 1990, to hear testimony on the safety of that blood supply. What the committee heard confirmed the headlines as witnesses testified that the Red Cross Blood Service was riddled with dangerous procedures, inept personnel, and a nonchalant attitude toward the importance of the process itself. What the committee also heard was a detailed history of the way that the blood industry in general, and the Red Cross in particular, had dragged their feet in the early days of the AIDS epidemic. Specifically, the blood bankers were accused of refusing to accept evidence, back in those early days, that AIDS could be transmitted by transfusion, and then downplaying the risk once they had admitted it existed. They were accused of resisting the use of the early blood tests that would have reduced the risk of contaminated blood. The were accused of making exaggerated claims about the ability of current tests to identify contaminated blood.

The committee also heard Dr. Marcus Conant, a professor at the University of California Medical Center, testify that in his estimation this foot-dragging had resulted in some twelve thousand to twenty thousand Americans being infected with the AIDS virus. "The blood bank industry," he told them, "is totally dependent on voluntary free donation of blood by altruistic citizens anxious to help their fellow man. While blood bankers do much good, it is also irrefutable that if donors do not come to blood centers there will be no product to sell to hospitals and patients. Blood bankers were terrified that if they questioned donors about high-risk behaviors, donors would cease to present themselves voluntarily to blood centers."[32] It was, he concluded, a decision motivated wholly by financial considerations.

The testimony before the subcommittee stung the Red Cross, and long after, Sandler showed a barely controlled indignation when he talked about the hearings:

There's no question in my mind that the public was confused by the sort of witnesses that were selected to testify. They were all highly critical, and not one of them made any mention of the success of the efforts by the ARC [American Red Cross] to reduce the risk of transfusion AIDS. I was greatly disappointed by the selection of the witnesses and the biases they represented. One was a commercial plasma distributor, one was a dermatologist who had no qualifications to discuss the subject, one was a hematologist who had a bias because of the infections that hemophiliacs have contracted. There were no qualified experts from the CDC, from the National Institutes of Health, from the ARC, from anywhere.[33]

In the wake of the embarrassment caused by the subcommittee hearings, the Red Cross knew that it needed a quick fix, and the man chosen to do the fixing was Dr. Jeffrey McCullough, one of the nation's leading transfusion medicine specialists. In August 1990, McCullough was made senior vice-president for Biomedical Services with a broad-reaching mandate from Red Cross President Elizabeth Dole to restructure the entire system, and in May 1991 he announced a drastic transformation plan that would:

- Temporarily close each Red Cross Center for eight weeks, during which time it would be reequipped and reorganized.
- Transfer the testing of blood from the fifty-four blood centers to fewer than ten regional laboratories.
- Install a single national computer system to record all test results.
- Expand patient services such as special typing, tissue services, and the return of a patient's blood during surgery.
- Separate the blood centers from the local Red Cross chapters, and allow them to function independently.

The Red Cross spent more than two years and $120 million in this transformation effort in the hope of creating a leaner, tighter, more efficient machine. But although some progress was made, that hope was not realized, and still is not. The Red Cross now claims that blood transfusion is responsible for only a tiny fraction of 1 percent of all the AIDS virus being spread in the country, and the FDA confirms that blood is acceptably safe at this time. Even Dingell's watchdogs agree that our blood supply has never been safer. But, as we shall see in chapter 12, the Red Cross has continued to be plagued by poor record keeping and a lack of quality assurance programs well into the 1900s.

Thus, the question remains. How safe is safe?

According to Dr. Gerald V. Quinnan, acting director of the FDA's Center for Biologics Evaluation and Research, "Whether blood is as safe as it 'should' be is a matter of judgment, and deals not only with technology assessment, but also with social choice and priorities. Blood, because it is a human biological tissue, is capable by its nature of transmitting disease. Yet the need for blood is most often related to a life-threatening condition, and a certain amount of risk is acceptable if it's part of a lifesaving treatment."[34]

The ultimate future of the blood supply may lie in the field of blood substitutes, artificial substances that perform some of the functions of human blood without the risk of transferring disease. One such substance is already available, a genetically engineered version of the human protein erythropoietin, which stimulates bone marrow to produce red blood cells, thus eliminating the need for transfusion in certain medical settings. Another product due to come on the market in the near future is a recombinant-derived clotting factor for hemophiliacs that will eliminate the use of the same factor derived from human plasma. But the ultimate breakthrough in blood substitutes is still over the horizon, a genetically engineered version of oxygen-bearing hemoglobin that would function as a true artificial blood. Optimism in this field is more theoretical than real, for the materials available at this time are too toxic to be used. Genetic engineering, however, offers the hope of eliminating the toxicities while keeping the oxygen-carrying abilities.

In recent years, the effort to make a blood substitute has turned to a technique known as "molecular farming," in which animals are the hosts for human genes. Using this technique, several chemicals, including a blood-clotting factor, have been made successfully, and most recently, DNX Inc., a biotechnology company in Princeton, New Jersey, has claimed to have developed a number of pigs that produce true human hemoglobin. The extracted hemoglobin has not been tried on human subjects, and may not succeed, but the company has given its research data to the FDA and plans to apply for approval for human trials. It's all a long way off, and until then we have to live with what we've got.

10

Eat It, It's Good for You!

THE PRO-INDUSTRY BIAS OF THE FDA DURING THE REAGAN-BUSH administrations was never more evident than in the area of food labeling. Responsible for regulating the truth of claims made on those labels, the agency during most of the 1980s demonstrated a distinct reluctance to antagonize industry, coupled with an attitude toward the needs of the consumer that came close to callous indifference. Some of this attitude was generated from within the FDA itself; more of it was forced on the agency from the White House through an Office of Management and Budget (OMB) that was fervently devoted to free-marketeering. But whatever the source, the FDA during those years conformed to the ethics of the time, which dictated that nothing must be allowed to get in the way of making a buck.

The buck was passed to the agency in 1984 when the Kellogg Company decided to challenge certain FDA labeling regulations by stating on its package that All-Bran cereal was helpful in the prevention of cancer. The claim was based on a generalized statement by the National Cancer Institute that, "A high-fiber, low-fat diet may reduce the risk of some kinds of cancer,"[1] but the claim also violated a decades-old FDA rule against making health claims for food. A product can be a food, or it can be a drug. Making a health claim makes it the latter, and makes it subject to the "new drug" and misbranding provisions of the law. The new Kellogg labeling was clearly improper, and the food industry as a whole watched anxiously to see what action the agency would take.

The action that the agency should have taken, but did not, rested on the complex and confusing responsibilities that the FDA bears for the safety and purity of the nation's food supply. The agency, while in charge of the regulations for most foods, has no control over meat and poultry, which is the concern of the U.S. Department of Agriculture (USDA). This division of responsibility serves to point up the different approaches to enforcement used by the two offices. The FDA is supposed to inspect food processing plants every two years, although that goal is rarely met in practice. The USDA, on the other hand, maintains continuous on-site inspections at meat and poultry processing plants, with the same inspectors virtually resident at the facility. There is much to be said for the USDA system, and the FDA, along with the public, would benefit from a routine of more frequent inspections; but the drawback to the USDA approach is that the same inspectors working day after day and side by side with the people they are inspecting tend to develop a "company" attitude. Thus, with the most honest of intentions and practices, the objectivity of the inspector is often eroded.

In another division of labor that seems self-defeating, the Fair Trade Commission (FTC) regulates the television and newspaper advertising claims made by food manufacturers, while the FDA regulates the labeling claims on packages in the stores. Thus, it is possible for the producer of a package of flour to make a claim for the product in a television commercial that would not be allowed in print on the package. The two agencies try to coordinate their efforts, but not always with success, and sometimes, as we shall see, with sharp controversy.

A third area of divisiveness lies in the regulation of foods imported from abroad. In this field the FDA is supposed to work in close cooperation with the Customs Service, but the liaison between the two agencies is often weak, and sometimes nonexistent.

The importation of foodstuffs is a particularly sensitive area for the FDA, with dangerously adulterated products reaching the American consumer mainly because the agency has neither the resources nor the authority to prevent such abuses. In 1988, the FDA estimated that some 40 billion pounds of food, at a value in excess of $20 billion, were imported into the United States. The FDA was able to check the paperwork on only 9 percent of those entries, and it physically sampled and tested slightly more than 2 percent.

According to a congressional report, of those entries that were physically tested, a staggering 40 percent failed to meet FDA standards for a variety of reasons, including contamination with bacteria, pesticides, insects, filth, and decomposition. The high rejection rate was a compliment to the agency's ability to focus its resources on high-risk imports, and the average violation rate was nowhere near that level, but it was a fair presumption that tens of millions of pounds of contaminated foodstuffs were nonetheless introduced into the American marketplace. The problem was not a small one, nor was it restricted to imported foods. The following year the Department of Agriculture estimated that in excess of 6.5 million Americans became sick each year because of microorganisms in their food, and some 9,000 afflicted persons died.[2]

A later FDA report identified several procedural improvements that would increase efficiency, but its basic conclusion was sobering:

> Given the expanding growth in the number of foodstuffs entering the United States from around the world, there can be little improvement in the efficiency of the present inspection program without additional resources and statuatory authority. Without additional resources, even additional efficiencies will not stop or

lower the rate of violative goods getting through the system, particularly in the largest import districts.[3]

Clearly, there were loopholes in the system big enough to drive a truck through, or a seagoing freighter, for that matter. One such case, and not a rare one, occurred in April 1988 when the San Francisco office of the FDA denied entry to a shipment of several imported products, including Coenzyme, Ginkgo Biloba, Milk Thistle, and Essential Fatty Acid. The products were being advertised as virtual cure-alls in the fields of cancer, heart disease, hepatitis, immunodysfunction, and hypertension; and they were rejected on the grounds that they were all unregistered new drugs. The goods, however, were reexported and then reentered into the country at Great Falls, Montana, and sold to a health food store before the evasion could be discovered. When the FDA finally realized what had happened, the goods were long gone.[4]

This case, and others like it, illustrated the need for the FDA to have the authority to embargo such products, and to seize them before they could be reexported and reentered elsewhere. It was a need that was recognized by Congress in a report by the Government Accounting Office (GAO) in 1984, which reviewed recalls and seizure actions by the agency in the fiscal years 1980–1982.[5] The report found that the seizure process took an average of sixty-five days, and that during the time that it took the FDA to obtain a court order, the agency had to rely on state regulatory authorities to detain the adulterated food, or hope that the importer would voluntarily hold the goods. The former was difficult, and the latter unlikely. State authorities did not always agree with the FDA, and even when they did it required valuable agency time and resources to get them to take action. As for the importers, the GAO report found that in nineteen out of seventy-six cases where the FDA had requested the firms to hold the products voluntarily, some or all of the adulterated food had been sold before the seizure action could be completed.

In theory, whenever an FDA-regulated product entered the

United States, the Customs Service forwarded the appropriate documents to the agency. If the entry was found to be adulterated, mislabeled, or otherwise violative by the FDA, Customs was supposed to supervise the destruction or the reexportation of the merchandise. And if goods were rejected, or if the importer had wrongfully distributed the goods before obtaining an FDA release, Customs was supposed to impose penalties, fines, and cause the importer's bond to be foreited.

In practice, Customs often failed to perform those tasks. Importers routinely filed false papers stating that rejected merchandise had been either destroyed or reexported, and then sold the same goods on the open market. A March 1988 study by the Import Operations Branch at FDA estimated that about 12.5 percent of rejected merchandise had been nonetheless distributed, while the New York District Office estimated that the number could have been twice that in their area.[6]

But if the liaison activities between the FDA and the Customs Service were inefficient to the point of failure, the real problem lay in the agency's lack of authority to enforce its own regulations. As a government study pointed out in the early 1980s, the power of a government agency to embargo or seize adulterated food was hardly a novel concept. The USDA had the power to detain or condemn adulterated food, stop production at a meat or poultry plant simply by withdrawing its inspectors, and hold up products from distribution for twenty days. The FDA had none of those powers. Further, the USDA had the statutory right to examine plant records, whereas the FDA did not. And the agency also lacked subpoena power to require the testimony of witnesses, or compel the production of documentary evidence. The FDA, in fact, was sadly lacking in many forms of legal muscle.[7]

But the most glaring lack was the inability of the agency to impose civil money penalties on those who violated the law. Violators could be prosecuted criminally, which sounded impressive, but the violations that were prosecuted were mostly misdemeanors. In fact, the FDA brought very few criminal cases then, mainly because of the glacial pace of the agency's review process described in chapter 2. Thus, a civil penalties regime would have been effective and appropriate for violations uncovered by the agency, and would have provided an active deterrent to further violations far more severe than the handful of criminal cases that were successfully prosecuted. Furthermore, the existence of civil penalties would have enabled the FDA to concentrate its limited criminal investigative abilities on those violations that truly deserved criminal prosecution. Clearly, the FDA in the 1980s was fighting a battle with the most basic weapons missing from its armory.[8]

But if the agency's lackluster performance in the field of food was generally due to an absence of resources and authority, its performance in the case of the Kellogg's All-Bran label was nothing less than an abdication of responsibility. The feeling inside the FDA about the Kellogg label was stated in a staff report saying that, "The wording on the Kellogg label . . . clearly ties the health-related information on the back panel to this one specific product, All-Bran, and is thus a brand-specific health claim of the type which we have consistently prohibited."[9] So strong was this feeling that FDA's Center for Food Safety and Applied Nutrition (CFSAN) and its Center for Drugs and Biologics (CDB) urged Commissioner Young to halt Kellogg's labeling campaign. In a November 1984 memorandum to Young, Dr. Sanford A. Miller, then director of CFSAN, wrote, "The time for FDA to act is now. Otherwise we will surely lose all control of health-related claims being made on food products. . . . Failure to act in the Kellogg case will be sending a signal to the entire food industry that FDA is abdicating its claim to regulate health-related claims on food labels."[10]

Along with this memorandum, Miller enclosed the draft of a regulatory letter for Young to send to Kellogg, requesting the company to terminate its All-Bran labeling campaign, but the letter was never sent. On January 19, 1985, Young advised the assistant secretary for health that the FDA was "preparing a notice that would announce a moratorium on label statements by Kellogg," but no such moratorium was ever imposed. The FDA had abdicated its responsibility, and as a result, one of Young's staff later observed that industry lawyers had begun to advise their clients to move ahead with health claims on food labels in light of the FDA's "having done nothing to Kellogg."[11]

On the surface, it is difficult to understand Young's lack of action. His position was strong, and his chief subordinates were outspoken in their desire to take action against Kellogg, but as in other instances, this commissioner could not bring himself to make the move. Testifying later before a congressional subcommittee, Young tried to shift the responsibility, at least in part, to NIH by protesting that the FDA had not been consulted in advance about the NCI health claim that Kellogg used on its label, and his protest was valid to a degree. NIH had no business approving a health-related claim for use on a food label, and by doing so it had poached on the territory of the FDA. But in a rather confused statement, Young showed how reluctant he was to criticize, even by implication, the action of another branch of the government.

"However, in looking at this myself, I felt that there was [sic] problems when we had the NCI recommending a course of action. Other portions of the Public Health Service were interested, and yet, what right would we have; how would we validate countervening claims and would it be able to hold up in court."[12]

It was a weak defense. It would be possible to make the case that Young was unwilling to declare the cereal an illegal drug because that would have meant seizing and impounding every box of All-Bran in the Kellogg warehouses, an anti-industry action of the sort that few agencies in a Republican administration would care to take (although a similar action would be taken several years later by a different commissioner in the Bush administration). But in actuality the decision was not Young's to make. Reagan administration officials from outside the agency played the pivotal roles in the decision-making process, particularly those from the Office of Management and Budget.

OMB's involvement in the matter derived from the authority granted to it under executive orders signed by President Reagan in February 1981. Those orders required the OMB to review all "major rules" that an "agency" intended to propose or issue. According to the order, agency regulatory action could not be undertaken unless the potential benefits to society outweighed potential costs.[13] Thus, in a very real sense, the OMB was regulating the regulators in an administration dedicated, at least in theory, to deregulation. And in such an administration, phrases such as "potential benefits" and "potential costs" tended to acquire meanings that were more ideological than rhetorical.

Thus, in the Reagan administration, a potential benefit to society was defined by its positive effect on industry-based revenues, and a potential cost by its negative impact. Given those definitions, it was incumbent on OMB to oppose any regulatory action that might limit the scope of the free market, while taking care not to invite a major public health disaster. Thus, it would have been naive to expect that OMB would adopt an approach to the question of health-claim labeling that was based on scientific objectivity. In the first place, there were no scientists at OMB, and in the second place, "objectivity" was another of those words that had taken on a meaning of its own.

So Young was far from being master in his own home when it came to health-claim labeling. He had pressures on him from several directions, as was witnessed by one OMB document, which stated that:

> The FDA Center for Food Safety and Applied Nutrition originally recommended that enforcement action be initiated against Kellogg. However, after a vigorous debate within the administration led by the FTC [Carol Crawford], OMB [Doug Ginsburg], and HHS [by then Assistant Secretary for Health Dr. Edward Brandt], FDA Commissioner Young was persuaded to resist entreaties of FDA career staff to take enforcement action against Kellogg in favor of a policy that the federal government should encourage, not discourage, manufacturers to provide verifiable health claims on food products.[14]

Leaving aside the suspect statement about "verifiable health claims," implicit in the administration position was the often-stated Republican claim that what is good for industry is good for the nation. But there was no good in the FDA decision, save what good would accrue to the benefit of the Kellogg Company, and in the ethos of the eighties that was quite good enough. Young caved in. Presented with the opportunity to make a decisive move in favor of the consumer, the FDA coughed politely, murmured a few words of disapproval and, like a well-bred lady faced with the unspeakable, turned its back and ignored the situation. The food industry relaxed, and the gates swung open to a new era in retail food merchandising.

Secure in the knowledge that the FDA was playing a hands-off game, Kellogg now made the same claim on its Cracklin' Oat Bran package as it had for All-Bran, referring again to the NCI statement about a "high-fiber, low-fat diet." But Cracklin' Oat Bran was not low in fat for a breakfast cereal. It contained 4 grams of fat per serving, raising the question of how a product that is not low in fat can make a health-related message about a high-fiber, low-fat diet.[15] The fat, in fact, came from unhealthy highly saturated tropical oils, but again the FDA took no enforcement action against the claim, and it was left to a private citizen, a heart-attack survivor named Phil Sokolof, to pay for a series of full-page advertisements in major newspapers condemning the practice. Along with Cracklin' Oat Bran, the ad pictured such products as Nabisco's Triscuits, Keebler Club Crackers, Quaker's Granola Bars, and Proctor & Gamble's Crisco. Once their sins were exposed, most of the companies reformulated their products, eliminating the tropical oils and leaving one to wonder at the sight of a single private individual doing the work of the FDA.[16]

But Kellogg was not ready to quit. The company had stumbled onto a merchandising mother lode by making health claims for its products, and next in line was Rice Krispies. This cereal had been around for decades in the same form, but now Kellogg began to promote the product as having "more energy-releasing B vitamins," while claiming on the back of the package that those B vitamins would produce a pick-me-up feeling in the morning. This was a technical half-truth, for B vitamins do help the body to convert fat, protein, and carbohydrate into energy at the cellular level, but one does not feel invigorated after eating a meal rich in B vitamins.

This sort of devious health claim soon became the norm, and before long a stroll down the aisles of a supermarket was akin to a walk around a drugstore. Labels and packages screamed their messages. Avoid heart attacks—eat olive oil. Strengthen your bones—drink calcium-fortified orange juice. Worried about cancer? Try our latest cereal. Previously hidden virtues were quickly found in old products, and hundreds of new products were introduced to fit every imaginable nutritional niche. By 1989, fully 30 percent of the $3.6 billion spent annually on food advertising included some form of health message.[17]

The hottest item was oat bran. The word *oat* or *bran* on a label was enough to turn virtually any product into a sure-fire seller, as General Mills found out when it rode the oat craze to a new height by turning Cheerios into the nation's most popular cereal—even though a 1-ounce serving contained only 1 gram of soluble fiber from oat bran. There were oat bran candy bars, oat bran potato chips, oat bran pita breads and bagels. The gullability of the American consumer seemed unlimited when a survey conducted by an advertising agency reported that 74 percent of the Pepsi drinkers interviewed said that they would switch to Coke if it had oat bran in it.[18]

Not to be outdone, Quaker Oats jumped onto the health-claim bandwagon by stating on its label that the product could "help reduce cholesterol when part of a fat-modified, low-cholesterol diet." Again, this was only part of the truth, and a very small part, at that. Quaker's own studies showed that a simple low-fat, low-cholesterol diet was twice as effective in reducing cholesterol levels as the oatmeal itself, and that the oatmeal provided only a slight additional benefit of borderline significance. The company also did not bother to inform the consumer that one would have to eat two bowls a day, seven days a week, fifty-two weeks a year, for every year of one's life in order to receive this minimal benefit.[19]

There were many ways to play this newest of marketing games. One was the portion-size ploy. Once it was clear that no wrists were going to be slapped, many canned-soup manufacturers changed the suggested serving sizes on their labels from 2 per can to 2.5 per can. Since the size of the can remained the same, the portions were actually reduced by 20 percent. That, in turn, reduced the numbers for fat, calories, and salt per portion, making the product appear to be more healthy although, in fact, nothing had changed.

"No cholesterol" was another label that soon began to appear on the supermarket shelves. With most consumers in total confusion over the meaning and function of terms such as saturated fat, unsaturated fat, and cholesterol, the label was slapped on such diverse products as bread, cookies, and lemons, even though those items, by definition, contained no cholesterol because they were vegetable products. Taking no chances, both Best Foods and Kraft introduced no-cholesterol mayonnaise, neglecting to advise the consumer that regular mayonnaise has little cholesterol to begin with.

But for all the concern over cholesterol and the race by manufacturers to distance themselves from the word, some researchers were begining to wonder how much the cholesterol that people ate in their food was directly linked to the cholesterol in their blood. Fergus M. Clydesdale, head of the food science department at the University of Massachusetts, pointed out that, "For the general population, saturated fat is a bigger issue."[20] Which meant that many consumers were playing with nutritional booby traps, for many of the products making the "no-cholesterol" pitch—cookies, crackers, potato chips—were high in saturated fat.

Another convenient device involved the word *light,* which, at that time, had little legal or regulatory meaning. It did, however, have a meaning to the consumer with its overtones of health and diet. At that point in the marketing frenzy, "light" or "lite" on a label had almost as magical an effect as "oat bran," or "low cholesterol." Riding the wave, two leading makers of olive oil, Bertolli and Filippo Berio, rushed to the market with products labeled "light." They were light, all right, but only in color and in taste, designed to address the dislike of many American consumers for the strong, natural taste of olive oil. The calories and the fats in the oil remained the same, but it was "light," and it sold.[21] Sara Lee tried it with a line of Light Classics desserts, in which the "light" referred to the airy texture of the products, and not to their caloric content. The FDA, of course, did nothing about it, but a group of state attorneys general threatened a legal battle and forced the company to change the name.[22]

By this time, consumer advocacy groups such as the Center for Science in the Public Interest were up in arms in protest against the half-truths and outright lies that were being foisted on the American buying public, and in indignation at an impotent FDA that was doing nothing about it. Of equal concern, they pointed out, was that while high-fiber cereals and low-fat dairy products were being pushed into the spotlight, no one seemed to be talking about the benefits of fresh vegetables and beans, products not backed by giant food packagers. Even a few industry executives seemed to be having second thoughts about using the health pitch to sell food. One of them confessed without apparent embarrassment that, "It's got to be here to stay if it sells products, but it's really a mixed bag because the consumer is so foolable."[23] And another, commenting on the absence of direction from the FDA, complained that, "Without guidelines, we almost run the risk of going back to the snake-oil era."[24]

While all this was going on, the FDA was not entirely insensitive to the plight of the consumer. In fact, the further history of the health-claims controversy was characterized by the heavy-handed actions of the OMB in overruling well-intentioned proposals by the agency. At this point, Young later testified, "FDA began to move more quickly towards developing the criteria that would underlie a new policy initiative in the area of health messages on food labels."[25] The result of this initiative was a policy statement approved by Young and forwarded to the secretary of HHS on March 17, 1986. The statement, in essence, maintained the traditional FDA position in that it did not call for any amendments to the agency's regulatory prohibitions against labeling food as effective in the mitigation, cure, prevention, or treatment of a disease.[26] The statement gave notice that the FDA, if given a free rein, would abandon its policy of nonenforcement, and now would uphold the law by regulating against health claims on foods. It endorsed the current prohibitions, stating that they "serve a critical public health protection function by safeguarding consumers from fraudulently promoted and sometimes dangerous products."[27]

This was strong meat to throw at the OMB, rather like trying to toss a beefsteak past a timberwolf. In May 1986 the statement was sent to OMB, where it disappeared into limbo for ten months, despite Young's request that it be given an expeditious review, finally surfacing at a meeting between the two agencies on March 23, 1987. The OMB's lack of action for almost a year reflected its dissatisfaction with the notice as being excessively restrictive to industry, as evidenced by an undated draft memorandum written by OMB desk officer Shannah Koss-McCallum.

"We have not duly acted on this proposal to date because we believe it unduly restricts claims beyond what is currently allowed in advertising."[28]

Thus, at its March 23, 1987, meeting with FDA, OMB stated that "some of the wording" in FDA's 1986 proposed policy statement was "overly restrictive" and stated its preference that FDA issue a proposed rule rather than a policy statement.[29]

What the OMB wanted was nothing less than the full revocation of the FDA's existing regulation that prohibited disease-specific health claims on food labels. The OMB felt that it could no longer count on the FDA to play ball. Young had played along by giving in on the Kellogg All-Bran case, but the regulatory zeal of his chief subordinates and the strong language of the 1986 policy statement indicated that the OMB might no longer be dealing with a bunch of happy campers.

The FDA could no longer be trusted, and Koss-McCallum wrote in a February 20, 1987, memorandum, "If we use a guideline approach [as the FDA had proposed], we would have to rely on the Food and Drug Administration's willingness to adopt a *no enforcement* policy [emphasis added]." And the OMB could no longer rely on that. The memorandum was chilling in its audacity. Here was a high official in the Reagan administration stating flat out that the FDA could no longer be considered as a trusted conspirator, could no longer be relied on to flout the law by not enforcing it. The OMB wanted a major role to play in the health claims policymaking process, a role that, Koss-McCallum wrote, "lets the marketplace, not the government, set the agenda for the types of claims that will be made" by providing the consumer with "information to decide whether the claim is objective or really just a sales pitch."[30]

This was too much for Young, who insisted that the OMB did not speak for the agency. Scientists, he said, not the lay public, would determine the validity of newly authorized, disease-specific health claims. But his indignation was largely wasted, for the compromise that then was hammered out between OMB and FDA satisfied no one. OMB did not get the rollback to the snake-oil era that it not-so-secretly desired, and the FDA did not get the full regulatory authority that it had requested. Missing from the 1987 proposal was the requirement that health claims be reviewed by Public Health Service agencies such as NCI. Missing was the requirement that claims be based on "a consensus of scientific opinion." Nor was it specified whether foods with nutritional drawbacks could carry disease-prevention claims (low cholesterol but high fat content), or whether claims would be limited to areas of the diet that posed true health problems for the average American.

This carpentered compromise drew heavy criticism from the public health, medical, and consumer communities, and the FDA, now sensitive to outside opinion, went back to the drawing board. This time it came up with a significantly improved regulation that would limit disease-prevention claims to five areas in which a consensus of scientific opinion supported a connection between diet and disease: sodium and heart disease; fats and heart disease; fats and cancer; fiber and cancer; calcium and osteoporosis. The FDA would undertake to develop model label messages and health summaries specifying how and in what contexts companies could use the messages. Manufacturers could still devise messages of their own, but these would be subject to FDA regulatory action.

It was a reasonable compromise, but the OMB would not buy it. In an effort at public persuasion, more than fifteen major organizations including the American Medical Association, the American Institute of Nutrition, the American Heart Association, the American Diabetes Association, the American College of Physicians, and the American Association of Retired Persons met with officials of the OMB in February 1989. They urged either that health claims on food labels be prohibited completely, or that the FDA's proposal be approved with the exception that food companies not be allowed to devise their own claims.

Bruce Silverglade, director of Legal Affairs at the Center for Science in the Public Interest, was at the meeting, and recalled that, "We were shocked by OMB's comments at the meeting. The officials present displayed a surprising ignorance of even basic public health, medical and consumer education principles." In the opinion of one official, "A regulatory policy that allowed some misleading and potentially dangerous disease-prevention claims might be justified if other helpful disease-prevention messages were encouraged by the policy."[31]

The discussions with OMB accomplished nothing. The Reagan White House was firmly committed to an ideology that placed the law of the jungle over the law of the land, that placed business interests first, and the interests of the consumer a distant second, that glorified individuality, but ignored the rights of the individual, that conceded casualties in return for profits. It was the lion and the lamb all over again, and the lion wasn't worried about cholesterol.

The result was an impasse. In July 1989, Dr. Fred Shank, then the FDA director of CFSAN, conceded publicly that due to the OMB's resistance, "this rule is going nowhere as currently written," and shortly after that Commissioner Young admitted that the agency had been forced by the OMB to go back to square one.[32]

One has to ask what more Young could have done at that point, and the answer is very little. The agency had made proposals that, although not fervently embraced by the consumer side, were sufficient to the moment. What was needed was a moratorium on disease-prevention claims until a proper regulatory framework could be worked out, but the OMB could not be moved on that point. The FDA was, then as now, a federal agency subject in the main to the wishes and even the whims of the party in power. It could propose, advocate, argue, and arbitrate a particular point of view; but if in the end that position was rejected there was nothing left but to submit. There was, of course, the alternative of honorable resignation, but that would have been a lot to expect from a creature of the bureaucracy. The nation was not in danger of another thalidomide disaster, or any sort of disaster, for that matter. A strict control over health-claim labeling on food packages was in the vital interests of the country, but it was not a cause over which one resigned.

So there was nothing for Young to do but to go along. The opportunity for action had been missed back in 1984 when the agency failed to take a stand against Kellogg in the All-Bran case. Sanford Miller had issued the warning then, when he had said, "The time for action is now." In 1984, a strong reaction to Kellogg's coup would have sent a firm message to industry that violations of the policy on health claims would not be tolerated. But a different message had been sent, the flood gates had been opened, and it was much too late to close them.

11

The Wunderkind

Any particular ten-year span need not define an era with mathematical precision. Most of the time, one decade slips casually into the next, unaware that its time is finished. Indifferent to labels, it leaves to the headline writers, the advertising agencies, and the poets the task of defining the age with a snappy title. The Gay Nineties at the turn of the century . . . the Depression years of the thirties . . . the insipid Eisenhower fifties . . . the Turmoil of the sixties—all were inexact labels. But there are times when the age and the calendar do coincide, or come close to it, as when the stock market crash of 1929 fell only a few months short of marking the end of the Roaring Twenties. In the same sense, the follies of the FDA, although certainly not limited to the 1980s, defined that decade for the agency, and the era ended with a mathematical neatness in December 1990 when Dr. David Kessler was selected as the new commissioner.

The beginning of the end was marked by the resignation of Dr. Frank Young in November 1989, forced from office by the uproar over the generic drug scandals. The author of some of the agency's more unfortunate adventures, Young was, in fact, a sincere and caring scientist and administrator whose main fault, according to one of his principal adversaries, was that "he wanted everybody to like him."[1] Even John Dingell had words of faint praise for Young's departure. "His intentions were good and he did his best when problems were brought to light," said the FDA's gadfly. "Under very difficult circumstances, Commissioner Young acted honorably."[2]

Upon his retirement, Young was made a deputy assistant secretary in the FDA's parent body, the Department of Health and Human Services, a technical promotion, but more of a face-saving move than anything else. He had served as commissioner for more than five years, longer than any other in the past quarter-century, and his most enduring achievement had been his belated attempt to respond to the AIDS crisis by proposing the new rules that eventually would give seriously ill patients early access to potentially lifesaving drugs.

But Young's term in office had been stained by a lack of vigilance in the area of consumer protection, and Sidney Wolfe, characteristically, could not let him go gently. "Frank Young is by far the worst FDA commissioner in the past eighteen years," said the agency's other gadfly. "Over and over again he has done what industry wants done as opposed to what medicine, science, and law wants done." And looking ahead to the selection of a new commissioner, Wolfe added, "There is nowhere to go but up."[3]

Young was replaced by Acting Commissioner James Benson, an agency veteran who, at the time of his elevation, had been head of the Center for Devices and Radiological Health. Benson was to serve as acting commissioner for one year as a search committee considered candidates, and no one envied him his job. He arrived on the scene just in time to take the heat for the sins of the past, with only the remotest chance of being selected as the permanent commissioner. For one thing, he lacked the proper degrees, and for another, he was too closely tied to traditional agency attitudes. The wheels were off the little red wagon at FDA, and the pressure from Congress, from consumer advocacy groups, from the AIDS community, and even from the White House, was for someone with a profound knowledge of food and drug law who could fix the wagon and get it rolling again.

The name that began to be heard was that of David Kessler, already known as a wunderkind with a medical degree from Harvard and a law degree from the University of Chicago. While in residency at Johns Hopkins, he had commuted between Baltimore and Washington, taking the night shifts at the hospital and working days writing food and drug legislation for Senator Orrin Hatch. In 1984, he became medical director of the hospital of the Albert Einstein College of Medicine in New York, and since 1986 he had been teaching food and drug law at the Columbia University School of Law. If any one person ever seemed fated through education and dedication to be FDA commissioner, it was David Kessler. He was thirty-nine years old when he took office.

Kessler knew that he was inheriting an organization in shambles, but he had no idea how far the situation had deteriorated. Even before he officially became commissioner he discovered, while trying to answer a routine question from Congress, that the FDA could not say how many regulations it had proposed over recent years, or what had happened to them. Only after agency staffers had searched through the *Federal Register,* through antiquated card files, and had sifted the memories of senior officials, was Kessler told that the FDA had proposed, but never had issued, about four hundred regulations. They ranged from the size of type required on drug packages to standards for lifesaving heart defibrillators.[4]

Peter Barton Hutt, a former FDA general counsel, commenting on the state of the agency when Kessler arrived, said, "I am absolutely astonished. How can you manage an agency if you don't even know what regulations you have proposed?"[5]

That was only the first of many bombshells that greeted the new commissioner. Another was the financial restrictions that bound the agency hand and foot. To pay for the job of regulating hundreds of billions of dollars of American and imported products, the FDA had an annual budget of $690 million, which was almost exactly the same as the budget of the community hospital in the Bronx that Kessler had just left. There would be other surprises to come, but two and one half years later, Kessler recalled that his first priority upon taking office had been agency morale. "I came to an agency that had been rocked by scandal and demoralized by the corrupt actions of a few, its resources diminished, its powers of enforcement vitiated by a tradition of casual laxity. I came to an agency peopled by dedicated professionals who had lost their sense of direction.[6]

"It was my job to provide a new direction, to win back the trust of the American people, and my first priority was to restore the agency's sense of pride. And I'm not talking about the sort of pride that you can wear on your sleeve for others to admire and applaud. I'm not talking about the pride that you can take in owning an expensive car, or a fine home. I'm not even talking about the justifiable pride of achievement in your chosen profession.[7]

"I'm talking about the private sort of pride that is always kept hidden. The pride that comes to your rescue during those dark times of the soul when doubts and disappointments go bump in the night. This is the sort of pride that we had to restore to the men and women who had been called to public service with the FDA. We gave them a cause in which to believe. We gave them a place where, once again, the good guys could win. We gave them the pride of accomplishment."[8]

Kessler brought his own brand of enthusiasm with him to the FDA, and it tended to be contagious. Louis Lasagna, a highly respected drug development expert at Tufts University, noted that, "I have never sensed in previous FDA commissioners—and I have known them all since the 1960s—the sense of purpose that Kessler exudes." George Gerstenberg, the FDA's Los Angeles district director, joined in the chorus. "The agency seems to be getting an esprit de corps. I see people putting in overtime when they didn't do that before. You see people staying late again. He has made a world of difference."[9]

But the corridors at Parklawn were not suddenly filled with

smiling faces singing "Happy Days Are Here Again." As bureaucrats, the men and women of the FDA knew that old habits die hard, and for some of them any old habit was preferable to anything new and challenging. Even the most optimistic of them, those who announced their willingness to work their butts off for the new commissioner, knew how difficult it would be to turn the agency around.

Kessler hit the ground running. He brought in new blood from the private sector to fill top-level jobs, and in a major reorganization he streamlined agency management by reducing to five senior officials the dozens of people who previously had reported to the commissioner. Within two months after taking over he had added a hundred new criminal investigators to the enforcement staff, many of them formerly with the Secret Service and the Drug Enforcement Agency, and had begun the process of whipping the enforcement procedure into shape.

In the old days, after a problem was discovered, an inspector's report had to clear fifteen levels of review before it reached the Justice Department (see chapter 2). Kessler trimmed the number of preliminary steps to five, enabling inspectors' reports to reach Justice within twenty-five days. In March 1992, he was able to say that since he took office the number of FDA prosecutions had risen from twenty-five in 1990 to forty-three in 1991, and that over the same period of time the number of injunctions had risen from nine to twenty-one.[10]

But these were all internal moves that were of little interest to the public. Kessler began to make headlines when, just two months after he had been sworn in, he stunned a gathering of food industry lawyers and executives by announcing that federal agents under his direction were about to seize thousands of cases of Proctor & Gamble's Citrus Hill orange juice. P&G's offense was the use of the word *fresh* on the Citrus Hill label, even though the juice was made from concentrate. Kessler said the label was misleading and wanted the offending word removed. P&G refused, confident that, as in the past, the FDA would back down or simply send a regulatory letter.

P&G was wrong. The U.S. Attorney's office in Minneapolis obtained an order from a federal magistrate allowing federal marshals to seize chilled and frozen versions of Citrus Hill's Fresh Choice orange juice stored in a Minnesota warehouse. That evening, a federal marshal and FDA inspectors hauled the juice away. Two days later, P&G agreed to drop the word *fresh* from the label.[11]

The action was hailed by consumer advocates such as Bruce Silverglade, who said that, "It shows that the federal watchdog is back on the beat. He's not just barking; he's biting."[12] And an op-ed story in the *New York Times* saw the move as a way to demystify food packaging and to clarify English usage. In an article titled "The Metaphysics of Orange Juice," Miles Orvell, a professor of English at Temple University, saw "a larger significance to the agency's action, for our food labels are as much cultural signs as signs of what's being sold. In challenging the claims of food manufacturers, the FDA, under Commissioner David A. Kessler, is reconnecting label and contents, word and thing."[13]

Kessler agreed, saying that, "The point about P&G was not so much about the one word *fresh* as it was about sending a message."[14]

He might have added that embedded in that message was a declaration that there was not going to be a repetition of the Kellogg's All-Bran case, when the FDA backed down. The FDA was back in business, and the agency struck again and again over the next few months. The word *fresh* was forced off the label of reconstituted spaghetti sauce, and then Kessler went after those ice cream and frozen yogurt makers who tried to present their products as being "fat-free." In a speech before the Center for Science in the Public Interest, he announced his intention to ban the use of misleading fat percentage claims on food packages.

"A product that says it is 97 percent fat-free but has 50 percent of its calories from fat is misleading," he insisted. "The ones that are the most misleading are the ones with the highest liquid content. This is a marketing gimmick and I'd like to see all of them gone."[15]

Kessler wanted nutrition labels to include comparisons between the daily recommended intake of fat and sodium and what a serving of a particular food provided. "Before I came to this agency," he confessed, "I did not carry around in my head that an appropriate daily intake on a 2,300-calorie diet was 75 grams of fat, or that 25 grams of saturated fat was in the ballpark. I didn't have any comparisons so the information on the nutrition panel was useless."[16]

At one point Kessler carried his campaign on nutrition labeling to the extreme of proposing special labels for children between the ages of six and twelve. The children's food label initiative was presented as a "personal brainstorm of Kessler, who, as a pediatrician and father of two young children, has a particular interest in reducing the amount of fat and salt in children's diets."[17] The proposal was opposed, understandably, by industry groups such as the Grocery Manufacturers of America, and supported by consumer groups, but never gained the support of the general public. Most people saw the concept as being too complex, too difficult to implement, and bordering on silliness. The *New York Times* agreed, commenting in an editorial on the faulty premise that children receive most of their dietary messages from packaging, and that they yearn to know what's actually in that cereal box—other than a prize. "But surveys reveal that most nutrition information is conveyed in school, at home and by doctors, books, TV, magazines and friends. . . . Dr. Kessler has neither the authority nor the support from industry to fiddle further with labels."[18]

It was time to retreat. With the best of intentions, Kessler had created a potential for ridicule, and he extricated himself from the situation neatly by handing the project over to a nonprofit information clearinghouse called Kidsnet, which promised to adopt and disseminate better nutrition information to children over the airwaves and in schools.

Over and over, Kessler repeated that he wanted to send a message both to industry and to the agency itself. During the Reagan years of regulatory decline the FDA had become, in

Kessler's words, "a large immovable object" incapable of action, and he wanted to teach the long-neglected field force to be decisive again. For industry, the message was equally clear. Referring to the agency's limited budget, he explained, "We are never going to have an FDA enforcement person on every plant in every corner. How are we going to enforce the act then? The only way I can deal with it is to create incentives [toward compliance] in spite of my lack of resources."[19] That incentive was fear. There might not be a cop on every corner, but there was one close by who wasn't about to look the other way. The days of casual enforcement were over.

Soft spoken and easygoing, Kessler was known to conduct meetings with his closest associates while curled up on a sofa with a ubiquitous can of diet soda by his side. But his public face was stern, and his manner driving. There was a feeling of swagger to this new commissioner, a sense of adventure that thrust him at once into the national spotlight. When he was teaching FDA law at Columbia, he often posed a hypothetical question to his students. "You're the commissioner of the FDA. It's four o'clock in the morning and you get a call saying that X product is contaminated. What would you do?"[20]

The call came for Kessler after only four months in office. It came in the middle of a weekend night from an aide in Washington State who told him that Sudafed capsules in the Seattle–Tacoma area had apparently been laced with cyanide, and so far two people were dead, and another was seriously ill. What do you do, Professor Kessler?

It would seem that an immediate recall of Sudafed twelve-hour cold capsules was the obvious course to take, but the question was more complex than it appeared. Should it be a partial recall, or a total? Also, previous recalls of poisoned products had pointed up the risks of consumer panic, imagined symptoms, and heavy losses to the manufacturer. "I had been preparing for that moment for years," Kessler said later. "There was no question in my mind."[21]

Even though the tampering appeared to be confined to a small part of the Seattle-Tacoma area, Kessler recommended an immediate nationwide recall of the capsules, and Burroughs Wellcome, the manufacturer, agreed. Eventually, six more boxes of capsules that had been poisoned were recovered, but there were no further deaths.

There had been occasions of tampering in the immediate past, notably the two Tylenol incidents, but Kessler's quick action in the Sudafed case caught the public attention. A cover story about Kessler and the "new" FDA in the *New York Times Magazine* was followed by television appearances, more magazine articles, and an intense coverage of agency activities by the communications media. Almost inevitably, the nicknames followed. First he was King Kessler, in recognition of his imperial manner, then David Howser, M.D., with reference to Doogie Howser, the precocious television teenage doctor, and finally Eliot Knessler, after the government agent who brought down Al Capone.[22] The image projected was that of a young and dedicated public servant intent on enforcing the law.

All of this enthusiasm for law enforcement was in direct contrast to the deregulatory spirit of the Reagan-Bush administrations, as the *Wall Street Journal* was quick to point out. "Some drug and food companies complain to the White House about the FDA chief's tough enforcement tactics on issues such as prescription drug promotion and food labels. Some conservative GOP backers contend he's a Democrat in Republican clothing. His growing reputation as 'Dr. E' [for enforcement] makes some Bush aides nervous."[23]

But despite the unhappy muttering coming out of industry, Kessler, during those early days, was perceived by the White House as more of an asset than a liability. With the Gulf War triumphs rapidly fading into the past, the Bush administration now seemed helpless in the face of the domestic problems that would eventually deny the president a second term. Kessler was high profile, and seen by the public as a straight shooter doing an honest job for the American consumer. As such, he was virtually untouchable, and in

August 1991, then White House chief of staff John Sununu told him that he was "the closest thing to a domestic policy" that the administration had.[24]

Actually, Kessler was as much a dedicated Republican as he was a practicing Tibetan monk. In fact, he had no political affiliation at all before entering government. But if the new commissioner was not a politician in the party sense, he had highly developed political skills in the art of the possible, and during the Bush years he used those skills whenever he had to.

Proud of his knowledge of inside Washington, he was far from the political maverick that he appeared to be, knowing when he could maintain his independent stance, but also knowing when to compromise. On two occasions, on the eve of appearances before congressional committees, he bowed to the wishes of OMB and canceled prepared testimony that the White House had found displeasing. But this was more surface than substance. The administration, while unwilling to have *its* FDA commissioner take a public stand in opposition to White House policy, was equally reluctant to oppose the agency publicly in the performance of its mandated duty. It was part of the price that had to be paid for fixing the little red wagon.

We have seen how, during the 1980s, a supine FDA and an OMB dedicated to the preservation of business interests opened the gates to a flood of questionable health claims for certain foods, which was soon followed by the irresponsible labeling of food products in general. The Nutrition Labeling and Education Act of 1990, sponsored by Representative Henry Waxman, was designed to close those gates and to provide the consumer with accurate and easy-to-understand labels on foods. Under the act, the FDA was given until May 1993 to standardize food labeling, particularly those labels that made implicit health and nutrition claims, such as "fat-free," or the use of words such as "light," or "lite," or "low cholesterol."

But Congress did not pass similar legislation to cover meat and poultry, which went into many processed food products, like microwave dinners and snack food. Thus, under the rules as they stood, a sausage pizza came under the jurisdiction of the Agriculture Department (USDA), which regulated pork, while a cheese pizza came under the guidance of the FDA, which covered dairy and most other foods. To complicate the situation further, the Fair Trade Commission (FTC), which regulated advertising, might well permit food companies to use terms in a television commercial that both the FDA and the USDA would not allow on food labels. Not since the days of the New Deal had there been such a wide-ranging effort to police the nation's food supply, and as a result the FDA would spend the next two years writing thousands of pages of regulations governing more than 14,000 foods.

Agriculture Secretary Edward Madigan promised to follow along on whatever regulations the FDA devised, but for an organization that was accustomed to thinking in terms of decades, rather than years, it was a daunting assignment for the FDA. At that point, no one realized the extent of the collision between the two agencies that lay at the end of the line.

Trouble appeared on the horizon, and resolution faltered at Parklawn, toward the end of 1991 when the agency cut a deal with OMB that weakened two significant aspects of the Nutrition Labeling and Education Act. The act stipulated that once the nutritional regulations were written and agreed upon, if a certain percentage of grocery stores nationwide did not voluntarily display the information, then mandatory disclosure would be required of all stores. The language of the act allowed the FDA to determine what the minimum percentage of stores would be, but directed that there should be a "substantial compliance."

The FDA proposed a minimum of 80 percent of chain stores and 50 percent of independents, but according to documents obtained by the Center for Science in the Public Interest, the OMB changed those figures to 60 percent of all grocery stores covered by law. The net effect was that 38,000 stores would be required to carry the information, rather than 45,000.[25]

And then there was the matter of "fruit drinks." By this time the nation's more sophisticated consumers were aware that in most cases a bottle of "apple drink," as opposed to "apple juice," contained only a small percentage of real juice. The same was true for other fruits. In order to control this merchandising maneuver, the FDA had proposed that the manufacturers of diluted "fruit drinks" use a standard procedure to determine the percentage of real juice in their product. This would be the figure given on the label, and would be used by the FDA in any enforcement actions. But the Bush White House, with its usual disregard for the consumer, wanted the manufacturers free to use any test they wished, meaning whatever test would give them the highest number for juice content.[26]

This prompted Representative John Conyers, Jr. (D–Mich.), chairman of the House Government Operations Committee, to comment that OMB's revisions appeared "to subvert congressional intent as expressed in laws to protect public health and safety." And, as if anyone had forgotten, Conyers noted that the OMB had, in the past, "forced the FDA" to weaken regulations governing health claims on foods.[27]

In its defense, the FDA issued a statement that had overtones of the eighties. "We don't think that the [OMB] changes are significant," said an agency spokesman. "They are a result of dialogue between the FDA and the OMB."[28] The dialogue, apparently, was more like a monologue, as the OMB changes prevailed.

Further weakening in the FDA position became apparent several months later when the agency published its proposals for making food labeling less confusing. The proposals, compiled under pressure from OMB, marked a distinct retreat from the FDA's original position. At issue was the standard to be used for allowing manufacturers to use claims such as "reduced fat" and "less sodium" on their packages. Kessler's original proposal sought to confine such use to products that really did incorporate substantial reductions. But his new approach, called "alternative," let the manufacturer qualify with only minor tinkering. The old and the new proposals looked like this:

In order to be labeled "reduced fat":

Under the original proposal, the product had to have at least 50 percent less fat than a comparable product.

Under the new proposal, the fat had to be reduced by more than 3 grams per serving, which was another way of saying only 3 grams per serving. This was actually no more than a minor reduction in something like a premium ice cream bar with 27 grams.

In order to be labeled "reduced calories":

Under the original proposal, the product had to have 33 percent fewer calories.

Under the new proposal, when compared with other standard brands, the product had to have more than (only) 40 fewer calories per serving. This was not much of a reduction on a 450-calorie serving of frozen cheese pizza.

In order to be labeled "reduced sodium":

Under the original proposal, the product had to have 50 percent less sodium.

Under the new proposal, the sodium had to be cut by a mere 140 milligrams, thus allowing the "reduced sodium" label on a salty can of soup that still contained 900 milligrams.[29]

FDA acceptance of the OMB guidance in these proposals represented a clear retreat from Kessler's original goals, made ironic in retrospect by his statement in June of the previous year about merchandising gimmickry in the labeling of ice cream. Reaction to these changes ran along ideological lines. Silverglade said that the action was "an eleventh-hour attempt to weaken food labeling reforms," while an official of the National Food Processors Association said with a straight face, "I think they [the FDA] are trying to make it clearer, and we are not opposed to that."[30]

The FDA about-face affected just one small part of the implementing regulations, but it illustrated what the agency had to contend with. Even with an action-oriented commissioner like Kessler in charge, the agency still was subject to the pressures of an industry-oriented administration. Kessler, willing to give away pieces to preserve the whole, saw the reversal as a political accommodation, and nothing more. In the end, it did nothing to tarnish his reputation, for the big battle was yet to come.

The battle that was joined pitted the FDA on one side and the USDA on the other, with the FTC dropping out of the contest for the moment. Kessler was content to have it that way. He had come into office believing that there should be no difference between the information on a package label and the manner in which the product is sold on television. But he now understood that hyperbole was at the essence of advertising, and, as he put it, "I don't expect that the Jolly Green Giant really exists."[31] Although he still felt that the FTC, if for no other reasons than those of uniformity, should adopt the FDA regulations, he was willing to put off that contest to another day and concentrate on his differences with the USDA.

The prime area of controversy between the two agencies was the back panel of the nutrition label. The FDA wanted to tell consumers not just how many grams of components such as fat or cholesterol a product contained but also to express that number as a percentage of the recommended total amount of fat or cholesterol that the average consumer should eat in a day. The assumption made by the FDA was that an average person consumed approximately 2,000 calories a day, including no more than 65 grams of fat.

USDA, on the other hand, argued against the listing of percentages, and held that labels should detail only the amount of each ingredient. USDA also argued that by basing the recommended daily values for fat and cholesterol on a standard of 2,000 calories, the FDA had made meat and poultry products—which tend to be higher in fat and cholesterol than vegetables and grains—seem unhealthier than they actually are for a large segment of the population. The way USDA saw it, 2,000 calories was appropriate only for women and children, and that many men could safely eat 2,500 calories daily.

The USDA action was an obvious violation of Secretary Madigan's promise to go along with whatever regulations the FDA might formulate. "Instead," commented a *New York Times* editorial, "he's chosen to lobby against proposed labels for foods over which he has no jurisdiction. If Congress wanted the nation's cattlemen and hog farmers to dictate health policy, it would have said so."[32]

What Congress had anticipated was just this sort of impasse, and it had included in the original legislation a provision called "the hammer," which established a deadline for agreement between the two agencies in early November 1992. If the deadline passed without agreement, down would come the hammer and into effect would go the original FDA proposals promulgated the year before. As the deadline approached, nobody really wanted the original proposals put into effect. They had been modified so often over the past year to accommodate the various interests involved that going back to square one would have satisfied no one. The final decision between FDA and USDA regulations lay with the White House, and with past experience as a guide, the conventional wisdom dictated that the administration would side with the USDA.

But the presidential election in early November changed all that. Bush lost, Clinton won, and the president was suddenly a leader with no more ideological axes to grind. For the first time in years, Bush was able to make a choice purely on merit, and the FDA regulations were clearly in the better interests of the nation. Reluctant to leave office labeled as a lame-duck obstructionist, Bush brokered a compromise that was, in effect, a victory for the FDA. With only minor changes, the agency's 4,000-page revision of food labeling regulations became law. Standing tall whenever he could, compromising whenever he had to, Kessler had brought a measure of sanity to the supermarketplace, and had changed the way Americans would eat for decades to come.

While Kessler's seizure of Proctor & Gamble's Citrus Hill orange juice served notice on the food industry that the FDA was back in business, his move against Syntex sent the same message to the pharmaceutical companies. According to the FDA, Syntex had made misleading claims for its product Naprosyn, an antiarthritis drug, and the agency demanded that the company mount a multi-million-dollar advertising campaign to retract those claims. When Syntex indicated a reluctance to do so, Kessler sent shock waves through the industry when he threatened to seize the company's entire $50 million inventory of Naprosyn if the FDA demands were not met.[33]

Syntex capitulated, the times were changing, and some things would never be the same again. Throughout the 1980s, those closest to the FDA had nervously observed the interaction of three factors. One was the agency's chronic shortage of working funds. Another was the ever-growing number of drugs being presented each year for approval. A third was the burgeoning biotechnology industry that was making new drugs available through genetic engineering. By 1992, the FDA was processing about 100 new drug applications annually, excluding generics, but there were about 300 more biotechnology products in the pipeline, and applications for their approval were expected in 1997. And the number of new biotech products was expected to rise to 1,200 or more a year by the end of the decade.[34]

By statute, the FDA was required to approve or disapprove a drug within six months of application, but it had been many years since that figure had had any meaning. In 1991 the mean time for approval of a drug had been 30.3 months, in 1990 it had been 27.7 months, and in 1989 it had been 32.5 months. With more and more applications to be reviewed, and with no increase in the number of people available to review them, there was approval gridlock waiting down the road unless the agency could find some way to speed up the process. The need was both humanitarian and financial. Important drugs were being kept from people who needed them, and, by FDA estimates, the pharmaceutical companies were losing an average of $10 million in potential sales for every month that a drug was kept from the market.[35]

There was an answer, and everyone knew what it was. It was an idea that had been proposed several times during the Reagan years. The plan was called "user fees," and through it the pharmaceutical companies would help to defray the costs of the approval process. It was, due to the wretched state of FDA finances and the refusal of the White House to improve that lot, the only device through which the agency could afford to hire enough examiners to work the heavy load of applications through the system.

The fee system would be broken down into three classifications:

1) A specific fee would be charged to the manufacturer for each new drug application submitted.
2) A flat annual fee would be charged to any manufacturer doing business with the FDA.
3) A flat fee would be charged for each product already on the market.

The system made sense, it was workable, but it never had been implemented. A high FDA official, speaking without attribution, explained why:

Under the "user fee" plan proposed by the Reagan people, any revenue that came in from drug companies would be used to offset funds provided for the FDA budget out of general revenue. In other words, if a million bucks came in from user fees, that would be one million less in the FDA budget. There would be no new money. You see, there are two ways that an antiregulatory administration can use to keep an agency like the FDA tied up. The first is to regulate the regulators, adding hurdles, layers of communications, and more avenues for review, binding them hand and foot. The second is to starve them, budget-wise. That was what OMB did during the Reagan-Bush years, and that was why the Reagan user fee plan was worthless to us. The drug companies were leery of it, too. They could see that it wouldn't work at all.[36]

Kessler had lobbied for a workable user fee plan from his first day in office, and by 1992, the situation was so clearly acute that the White House had to take notice of it. It was, after all, an election year, and in addition to their own constituents, the large pharmaceutical companies were crying for some form of action. Reluctantly, the Bush people let it be known that they would consider some form of user fee plan that would give the FDA the funds it needed to expedite the review process. An act of Congress was required, and on the Hill the bill was drafted by the staffs of Dingell and Waxman in the House, and Kennedy and Hatch in the Senate.

The process was not without negotiation. The fees to be charged had to be adequate, but not ruinous, and appropriate to both large and small manufacturers. Furthermore, a standard of performance had to be established for the FDA, ensuring that the money spent would produce the desired results. In the end, the Prescription Drug User Fee Act of 1992 looked like this:

1) A fee of $100,000 for each application, rising to $233,000 per application in five years.
2) An annual fee of $50,000, rising to $138,000 in five years.
3) A fee of $6,000 annually for each product already on the market, rising to $14,000 in five years.

In return, the FDA pledged to hire six hundred new examiners, and to cut the review time from the current twenty months to six months for the most important drugs and twelve months for others. The law would expire in five years, so that if the agency failed to make the goals, it could be held accountable when a new law came to be drafted. The measure was expected to raise more than $300 million over that same five-year period.[37]

Passed by the Congress and signed into law, the user fee bill was greeted with almost universal approval. Some dissenters expressed the fear that the act would "blur the boundaries between the agency and the industry it is charged with regulating," but

Congress, industry, and the advocacy groups were content with what they had wrought.

"This is a major turning point—it's a milestone," said Kessler. "In the end, it may be the most important thing we have done."[38]

In the end, it was a long step away from the agency's decade of folly, with folly described in Tuchman's sense as a conspiracy against one's own best interests. Ideologically driven to minimize the effect of the regulatory agencies on the marketplace, Reagan-Bush had come close to jamming the gears of the FDA, and during the 1980s the agency had accepted the prospect with a regrettable equanimity. Now a step had been taken in the other direction, and as important as its practical goals might have been, the step was equally important as a sign of the agency's retreat from folly.

12

Commencement Day

SHORTLY AFTER DAVID KESSLER WAS NAMED COMMISSIONER OF THE FDA, a
writer sat in the office of the chief executive of one of the largest
pharmaceutical manufacturers in the United States. The office was
grand, richly furnished, and impressive paintings were hung on the
walls. The writer was there as a working journalist, but he found
himself treated as an honored guest, seated with his host in deep
armchairs while coffee was served to them in shells of fragile china
by a white-gloved servant. The writer later learned that the servant
was his host's driver, but that he put on the white gloves whenever
there was coffee to be served. It was very good coffee, of a quality
that the writer rarely had a chance to sample. He knew what fine
coffee smelled and tasted like, and this was it. This was the kind of
coffee that you didn't import from Colombia—you smuggled it in.

For some irrational reason, the excellent coffee offended the
writer. He felt that he was being softened up, his professional
objectivity threatened; but the feeling passed quickly. This was
nothing more than the coffee that his host drank every day. He was
accustomed to it, and the writer wasn't. People who live at the top
of the industrial heap, with an original Modigliani on the wall, do
not have to seduce journalists with high-class java.

Host and guest chatted for a while about the new man running
the FDA, and the writer offered the observation that Kessler
seemed to mean it when he talked about being an "enforcement"
commissioner. The executive allowed himself a faint smile. He
moved his head fractionally from side to side, and murmured, "Do
you really think so? Well, we'll see. We'll see. *Plus ça change, plus
c'est la même chose.*"[1]

The writer thought about those words slightly more than two
years later as he wandered over the campus of a prestigious
southern university on a springtime afternoon. He was there to
hear David Kessler deliver a commencement address. The
commissioner was much in demand now for that sort of thing. But
before the speech the writer needed a cup of coffee. He needed
the coffee badly. The night before had been a late one, and he had
risen equally late and missed breakfast. He saw a refreshment stand
on the far side of a stretch of manicured lawn, close to the stadium
where the commencement exercises would be held, and he
headed in that direction. What he wanted was a cup of coffee on a
par with the excellent brew that had been served to him two years
before, but he knew that he was not going to get it. He had been
searching for that perfect cup of coffee for two years now, he had
yet to find it, and he was close to the conclusion that what he had
been served that day was reserved for those who dwell in high
places.

Memories of coffee fueled his thoughts and, considering all the
changes that had gone on at the FDA since the day of that perfect
cup, he wondered if his host of two years before still thought that
the more things changed, the more they stayed the same. He
decided that the man probably did. People who express themselves
in French clichés do not change their minds easily.

But there had been changes, many of them. Changes in the way
that the law is enforced, changes in the way that new drugs are
approved, changes in the way that foods are marketed, changes in
the way that the markets are regulated.

And that is the nut of the problem, he told himself. Always has been, and always will be the primary question that faces any regulatory agency. How much do you regulate, and when?

Arthur Schlesinger noted in his preface to William Patrick's book on the FDA that the question traces itself back to the days of the Founding Fathers of our country. "The men who gathered in Philadelphia in 1787 to write the Constitution had two purposes in mind. One was to establish a strong central authority, and the other was to prevent that authority from being abused. If men were angels, no government would be necessary, said the 51st Federalist Paper. But government was necessary, then as now, because angels are always in short supply."[2]

Strolling along, badly in need of a coffee fix, the writer came to the something-less-than-profound understanding that although government can be obstructive and costly, it is also essential. Yet Americans have always regarded their government with a mixture of reliance and mistrust. It could almost be called the American way. When pollsters ask Americans large and generalized questions—"Do you think that government is too involved in your life? . . . Do you think that government should stop regulating business?"—a majority of the people will always vote to get the government off their backs. But when specific questions are asked—"Are you in favor of Social Security? . . . How do you feel about Medicare? . . . What about the safety of your food and drugs?"—the same majority will vote for government intervention.

In fact, Americans do not want less government. What they want is a government that can solve problems, is more efficient, and is more capable of making changes.

Yes, there had been changes at the FDA. The generic drug scandal was only an ugly memory now, with a new man, Roger Williams, running that department. In the past, the agency would inspect a generic plant on a lackadaisical schedule, every few years. Now, when an application was submitted, the agency inspected the plant before approval was given to be sure that the manufacturer could do what it said it would do. It wasn't a guarantee against corruption, but it made sure that legitimate players were in the game.

There was a new head, too, at the Center for Devices and Radiological Health, Bruce Burlington, who had helped clean up the generic drugs mess. Kessler had ordered a review of the center's approval process for devices, and the report of the committee, chaired by Robert Temple, showed how much of a need there was for improvement. Four major areas of reform were put into place.

> *Expedited review.* Under the new system, CDRH would separate products that offered innovation in health care from those that merely copied existing products.
>
> *Risk assessment.* Under the old system, every product under consideration got the same general kind of review. Under the new system, uncomplicated products such as adjustable headrests for dentists' chairs would receive only administrative review. As Burlington pointed out, "It doesn't deserve a team of rocket scientists to work that out."[3]
>
> *Status reports.* Under the new system, companies would be able to track the status of their applications.
>
> *More stringent filing requirements.* Under the new rules, the FDA would no longer accept vague or incomplete applications, but would return them at once for prompt revision.

This was progress of a sort, a point which even the Dingell subcommittee acknowledged, while at the same time ridiculing the fast-track approval of devices of questionable merit. The subcommittee noted that, "The determination made by the FDA that a 'female condom' is substantially equivalent to a 'male condom' could have been refuted by any electrician who ever tried to interchange a 'male' with a 'female' electric plug."[4]

Still, the subcommittee noted that, "The FDA's willingness to recognize its past failings, and to undertake significant reforms, is very encouraging."[5]

Also encouraging to many were the changes in the control of silicone breast implants. In April 1992, after a year of review, the agency announced that the implants would be made available only for reconstructive surgery. Women who sought them for purely cosmetic purposes would receive them only if they qualified for small, tightly controlled clinical studies. Movement, but not enough, said the critics. The decision was supposed to set in motion a major study of the effects of implants on the health of the recipients, but a report from a House Government Operations subcommittee charged that the agency had not followed up sufficiently on those plans. It said that data collection was lagging, and that the documentation from doctors was being sent not to the FDA but back to the manufacturers. Shades of the "self-policing" approach that had caused the problem in the first place.

The FDA contested the report, and Kessler pointed out that the highly publicized hearings on the subject had had their own effect by sharply reducing the number of implants performed: only 3,500 in 1992 compared to a previous average of 130,000 a year. But a *Washington Post* editorial pointed out that, "Raising awareness is a different matter from enforcing compliance . . . especially when lax compliance was what created the problem in the first place."[6]

But then, in a more positive move, lawyers representing both the defendants and the plaintiffs in the many lawsuits over silicone breast implants announced a proposed $4.75 billion fund to compensate women with the implants. Under the plan, a woman with an implant who had acquired one of eight specific diseases would collect $200,000 to $2 million, and others could seek lesser amounts to pay for the cost of removing implants and for other medical care.

The writer followed the crowd that was heading toward the football stadium where the commencement ceremonies would take place, and where Kessler would speak on the subject of public service and public servants. Still fuzzy from a lack of both sleep and coffee, the writer's mind was filled with soft and fluffy concepts of justice and compensation. How much money does it take to compensate for decades of greed and stupidity? Who won and who lost? Who were the good guys and who were the bad guys?

With thoughts such as those, it was definitely time for some coffee, but the coffee, when purchased at the refreshment stand, was a predicatable disappointment. The writer took a couple of sips, and left the plastic container on the counter. His Proustian search for that memorable cup would go on. The more coffee changed, the more it stayed the same.

There had been changes, too, in the way that drugs for AIDS and cancer were being approved, philosophical changes. After AZT came the drug ddI, and when it was presented for approval Kessler told the advisory committee that, "Our goal is to turn protocols around in weeks and to measure review times in months."[7] The data submitted were anything but complete, and a far cry from what would normally have been required, but Kessler broke with the traditional FDA conservatism and told the committee that it was time to take risks. "I said that it was okay if they were wrong in the end. I was dealing with a bunch of scientists who, by their very nature, would look at something and give you all the reasons why it wasn't going to work. I wanted to convince them that it was all right to take a risk."[8] The committee approved ddI by a vote of 5-2, with one abstention.

Change in the way that the nation's blood supply is protected. In 1988, the FDA and the American Red Cross had entered into a voluntary agreement that governed oversight of the blood program, but despite the ARC's ambitious overhaul plans the violations had continued. What then ensued was a running internal dispute that pitted the FDA field force against the FDA center in Bethesda. The field wanted action taken, while the center, with the superior air that every home office tends to adopt, felt that it could reason with its old friends at ARC. In the end, the field won, which in itself was a refreshing change from the old days. Kessler took the ARC into court and got a consent decree over how the Red Cross ran its blood collection operations, thus replacing the voluntary 1988 agreement. The result: a more comprehensive quality assurance program, increased training for blood program workers, and improved data and records management.

"There is no single way to protect blood in this country," Kessler had explained. "The system is built on a series of protective layers, and no one layer picks up everything. You defer donors, you do the tests, and you go back to check the records and see if an infected person has donated before. If the blood banker violates one of the layers you don't necessarily have a contaminated unit, but our standards insist that each layer has to be intact."[9]

The problem resided in the nature of blood banking. In the case of a pharmaceutical manufacturer with repeated inspection violations, the FDA can close it down and seize its product. But the Red Cross provides 50 percent of the nation's blood supply. Clearly, the FDA cannot close it down, and cannot seize its product. For any patient requiring a blood transfusion, the risk of not receiving that transfusion outweighs the risk of receiving blood.

The crowd that was streaming toward the stadium was festive without being noisy. A certain solemnity went with the occasion. Family, all of them, filled with the pride of accomplishment and the knowledge that, for most of them, there would be no more tuition bills to pay. The writer followed the crowd into the stadium and took a seat in a rear row.

More changes? There was MEDWatch, a new program designed to encourage health professionals to report serious side effects or defects in a broad range of medications and materials, from drugs and devices to infant formulas and dietary supplements. In the past, physicians had not consistently reported adverse effects, if they recognized them at all, and by some estimates as many as 13 percent of hospital admissions could be traced to bad reactions to drugs or devices.

"The reporting of serious adverse events has to become part of the culture of medicine,"[10] Kessler had said, and Sidney Wolfe, in applauding the new program, had predicted that, "If we double the number of reports we get, it would literally take half as long to get the information needed to take action."[11]

So, the retreat from folly had begun, but there was so much more to be accomplished, and the writer had to wonder if it could be done. Realistically, was there any power on earth that could turn a bureaucratic behemoth into a sleek, well-oiled machine? That was the point that troubled the writer, and the point that Kessler would address today. Public service and the public servant. It troubled the writer because he was not a believer. Sitting there with the sun on his face, and a drowsiness within him that could only be dispelled by a cup of decent coffee, he confessed to himself his dirty little secret. Which was that he had no faith in the civil service system. No faith in the public servant. No faith in anyone who could not be fired for doing bad work. Like the people who had screwed up on the so-called killer grapes. Not only had they kept their jobs, but they had all been given an FDA award of merit. Why?

"Morale," said a high FDA official anonymously after the writer had published the story. "You had the facts exactly right, but . . . morale, you know?"[12]

No, he didn't know. Why was the harassment of Barr continued long after it was clear that the company was guilty of little more than whistle-blowing?

"Morale," was the answer. "You have to stand behind your people, you know?"[13]

No, he still didn't know. He knew that the alternative to civil service was a great deal worse than tenured incompetence, but he could not applaud a system that not only condoned failure but protected it.

All right, there was nothing rare about that—many people felt that way—but the writer was not comfortable with the feeling. He wanted to believe, but he couldn't. In the world in which he lived, it was a tenet of faith that only those who lacked the wit and the drive to make it in the real world fled to the shelter of public service. And yet, objectively, he knew that there were many hardworking, intelligent—even brilliant—career public servants in the FDA. He had met them, interviewed them, been charmed and impressed by their drive and their knowledge. Why, then, could he not believe that a few of those people, led by a dedicated activist like Kessler, would eventually make a difference at the agency?

At that point, Kessler was introduced, and he began to speak. The writer tried to pay attention to what he was saying, but his thoughts wandered. Too little sleep the night before, not enough coffee, and a hot sun accented his drowsiness. The commissioner was talking about the need for public service to attract the best and the brightest, and he heard phrases like *people of judgment . . . people of courage . . . people of dedication . . . people willing to serve the national interest, not the special interests.*

He was speaking of a public service that everyone deserves, but rarely ever gets. Conventional fare for a commencement day address, but all the writer could think of was how close the nation had come to a series of public health disasters in the eighties. How close we had come to the breakdown of a system designed to protect us. How far the FDA had strayed from original intent. And all in the name of an ideology that was nothing more than the law of the jungle. Nothing more than greed disguised as enterprise. Nothing more ennobling than the work of a thief in the night.

And now he was being asked to believe that a new day was dawning. And it had. He had seen the sunshine, seen the rays of change. He had seen a new and active leadership at the highest levels working for that change. At the highest levels, yes, but what about the body of the beast? What about the mass in the middle, those whose jobs were ever secure? Had they also seen the light? Had the passions and the prejudices that had driven them in the past been suddenly transformed by the sunlight into pure reason? Would that it were so, but he found it difficult to believe.

Kessler came to the end of his carefully crafted speech. There was applause. The writer stood up, and found that he was applauding, too. He was applauding the man's vigor. He was applauding his intentions. But could he make it work? The writer certainly hoped so. Still suspicious, still doubtful, still hoping, he went looking once again for that perfect cup of coffee.

NOTES

Chapter 1: A Flock of Rejected Suitors

1. *Final Report of the Advisory Committee on the Food and Drug Administration.* U.S. Department of Health and Human Services, Washington, D.C., May 1991.
2. Wolfe, Sidney. *Public Citizen Health Research Group Position Paper 1154,* January 17, 1989, Washington, D.C., p. 1.
3. Janssen, Wallace F. *The U.S. Food and Drug Law: How It Came, How It Works.* U.S. Department of Health and Human Services Publication (FDA) 86-1054, Washington, D.C.
4. Holmes, Oliver Wendell. *Medical Essays, 1842–1882,* 2nd ed. Houghton Mifflin, Boston, 1883, pp. 202–203.
5. Wolfe, Sidney. *Public Citizen Health Research Group Position Paper 869,* November 18, 1982, Washington, D.C., p. 1.
6. Zamula, Evelyn. "Reye's Syndrome, The Decline of a Disease." *FDA Consumer,* November 1990, p. 21.
7. Wolfe, Sidney. *Public Citizen Health Research Group Position Paper 836,* May 24, 1982, Washington, D.C., p. 4.
8. Wolfe, Sidney. *Public Citizen Health Research Group Position Paper 917,* November 7, 1983, Washington, D.C., p. 1.
9. Wolfe, Sidney. *Public Citizen Health Research Group Position Paper 922,* November 29, 1983, Washington, D.C., p. 1.
10. Ibid.
11. Wolfe, Sidney. *Public Citizen Health Research Group Position*

Paper 982, November 5, 1984, Washington, D.C., p. 1.

12. Wolfe, Sidney. *Public Citizen Health Research Group Position Paper 1005,* March 15, 1985, Washington, D.C., p. 1.

13. Tuchman, Barbara W. *The March of Folly.* Alfred A. Knopf, New York, 1984, p. 4.

Chapter 2: A Clear and Urgent Need

1. David W. Nelson, former chief investigator for the House Subcommittee on Oversight and Investigations for the Committee on Energy and Commerce, personal communication.

2. Unpublished memorandum issued by U.S. House of Representatives, Committee on Energy and Commerce, Subcommittee on Oversight and Investigations, Room 2125, Rayburn House Office Building, Washington, D.C. March 6, 1991, p. 3.

3. David Nelson, personal communication.

4. Thottam, Jyot. "Generic Drug Makers Prepare for Their Next Battle." *Wall Street Journal,* August 9, 1993, p. B4.

5. David Nelson, personal communication.

6. Teitelman, Robert. *Gene Dreams.* Basic Books, New York, 1989, p. 15.

7. Blair, John. *Economic Concentration.* Harcourt Brace Jovanovich, New York, 1972, pp. 496–497.

8. Ibid.

9. Ibid., p. 401.

10. Janssen, Wallace F. *Oral History of the Food and Drug Administration.* Accession 436. History of Medicine Division, National Library of Medicine, Bethesda, Md. January 30–31, 1984, p. 78.

11. Teitelman. *Gene Dreams,* p. 141.

12. *Prescription Drugs.* General Accounting Office Publication, GAO/HRD-92-128, August 1992, pp. 3–4.

Chapter 3: The Man in the Coonskin Cap

1. Silverman, Milton, and Philip R. Lee. *Pills, Profits, and Politics.* University of California Press, Berkeley, 1974, p. 111.

2. Ibid.

3. Ibid., p. 112.

4. Ibid.

5. Silverman, Morton. *The Drugging of the Americas.* University of California Press, Berkeley, 1976, p. 118.

6. Rankin, Winton R. *Oral History of the FDA.* OH82. History of Medicine Division, National Library of Medicine, Bethesda, Md. September 30, 1980, p. 28.

7. Ibid., p. 29.

8. Ibid., p. 32.

9. Silverman. *Pills,* p. 11.

10. Rankin. *Oral,* p. 32.

11. Schlesinger, Arthur M. *A Thousand Days.* Houghton Mifflin, Boston, 1965, p. 722.

12. Rankin. *Oral,* p. 33.

13. Reeves, Thomas C. *A Question of Character.* Free Press, New York, 1991, p. 331.

14. Rankin. *Oral,* p. 37.

Chapter 4: A Mess in Rockville

1. McCombs, Phil. "The Bungled Punishment of Prisoner Seife." *Washington Post,* April 3, 1992, p. A1.

2. *FDA'S Generic Drug Approval Process. Part Two.* Hearings before the Subcommittee on Oversight and Investigations of the Committee on Energy and Commerce, House of Representatives, Serial No. 101-115. U.S. Government Printing Office, Washington, D.C., 1990, p. 5.

3. Ibid., p. 12.

4. Ibid.

5. Ibid., p. 13.

6. Ibid.

7. Ibid., p. 14.

8. Ibid., p. 18.

9. Ibid., p. 14.

10. Ibid., pp. 16-41.

11. McCombs. "Bungled Punishment," p. A1.

12. Ibid.

13. Ibid.

14. Ibid.

15. Ibid.

16. Ibid.

17. Hearings before the Subcommittee on Oversight and Investigations of the Committee on Energy and Commerce, House of Representatives, Serial No. 101-163, May 3, 10, and 11, 1989. U.S. Government Printing Office, Washington, D.C., 1989, p. 125.

18. Ibid., pp. 125–126.
19. Ibid., p. 127.
20. Ibid.
21. Ibid.
22. Ibid.
23. Freudenheim, Milt. "Market Place." *New York Times,* June 16, 1992, p. D10.
24. *Transcript of Motions.* U.S. District Court, District of Columbia, Docket #CA88-3402, January 11, 1993.
25. *Prescription Drugs.* General Accounting Office Publication, U.S. Government Printing Office, Washington, D.C., GAO/HRD-92-128, August 1992, p. 4.
26. Sawaya, Zina. "Getting Even." *Forbes,* April 29, 1991, p. 92.
27. *Fairness in the Food and Drug Administration's Generic Drug Program.* A consultants' review by Kibbe, Arthur H.; Kopf, James A.; and Zarembo, John E. Front papers not numbered. Cited material appears under conclusions by individual members.
28. Greenfield, Jerome. *Wilhelm Reich vs. the U.S.A.* Norton, New York, 1974, pp. 91–92.
29. Freudenheim. "Market Place."
30. Tom Dornay, personal communication.
31. *FDA Bioequivalence Task Force Report.* January 1988, p. 5.
32. *Washington Post,* "Ex-Generic Drug Maker," January 23, 1993, p. A2.
33. McCombs. "Bungled Punishment."

Chapter 5: Toys 'R Us

1. *The New Our Bodies, Ourselves.* The Boston Women's Health Book Collective. Simon & Schuster, New York, 1992, p. 249.
2. Sobol, Richard B. *Bending the Law.* University of Chicago Press, Chicago, 1991, p. 1.
3. *New Our Bodies,* p. 250.
4. Sobol. *Bending,* p. 2.
5. Ibid., p. 5.
6. Ibid., p. 8.
7. Ibid., p. 9.
8. Ibid., p. 11.
9. Ibid., p. 13.
10. Ibid., p. 10.
11. Ibid., pp. 340–342.
12. Tony Luizzo, personal communication.

13. Theresa Luizzo, personal communication.

14. Ibid.

15. Sidney Wolfe, personal communication.

16. Transcript of Hearing before the Subcommittee on Oversight and Investigations of the Committee on Energy and Commerce, House of Representatives, Serial No. 101–127. U.S. Government Printing Office, Washington, D.C., 1990, p. 270.

17. Sidney Wolfe, personal communication.

18. Ibid.

19. Wolfe, Sidney. *Public Citizen Health Research Group Position Paper 967,* Washington, D.C., p. 2.

20. Ibid., pp. 2–3.

21. Ibid., p. 3.

22. Ibid.

23. Ibid., p. 1.

24. Sidney Wolfe, personal communication.

25. Bruce Finzen, personal communication.

26. *New York Times,* "Lawsuit Settled Over Heart Valve," January 25, 1992, p. A9.

27. *New York Times,* December 12, 1991, p. D1.

28. *Associated Press Bulletin.*

29. *New York Times,* "Company Confident on Heart Valve Costs," November 12, 1992, p. A20.

30. Tony Luizzo, personal communication.

31. Hilts, Philip J. "Strange History of Silicone." *New York Times,* January 18, 1992, p. A1.

32. Ibid.

33. Ibid.

34. *New York Times,* editorial, October 25, 1991, p. A32.

35. Hilts. "Strange History."

36. Hilts, Philip J. "Implant Maker Is Said to Shun Better Designs." *New York Times,* January 15, 1992, p. A18.

37. Hilts. "Strange History."

38. Ibid.

39. *Wall Street Journal,* "Breast Implant Debate," February 14, 1992, p. A1.

40. Rensberger, Boyce. "Breast Implant Records." *Washington Post,* November 3, 1992, p. A3.

41. Ibid.

42. Gladwell, Malcolm. "Documents Tell Risks of Implants." *Washington Post,* February 11, 1992, p. A1.

43. Ibid.

44. Ibid.

45. Woods, John E., and Phillip G. Arnold. "Fiction Obscures the Facts of Breast Implants." *Wall Street Journal,* April 7, 1992, p. A16.

46. Burton, Thomas M. "How Industrial Foam Came to Be Employed in Breast Implants." *Wall Street Journal,* March 25, 1992, p. A1.

47. *Washington Post,* "Promising Medical Devices to Be Speeded to Market." June 25, 1993, p. A2.

Chapter 6: Vitamins, Hurricanes, and Killer Grapes

1. Williams, Lena. "FDA Steps Up Effort to Control Vitamin Claims." *New York Times,* August 9, 1992, p. A1.

2. Williams, Lena. "A Correction: No Plan to Classify High-Potency Vitamins as Drugs." *New York Times,* August 16, 1992, p. A33.

3. Toufexis, Anastasia. "The New Scoop on Vitamins." *Time,* April 6, 1992, p. 54.

4. *New York Times,* letter to the editor from Sen. Orrin G. Hatch. October 27, 1992, p. A22.

5. J. B. Cordaro, president of the Council for Responsible Nutrition, personal communication.

6. Toufexis. "The New Scoop," p. 57.

7. Hausman, Patricia. *The Right Dose.* Rodale Press, Emmaus, Pa., 1987, p. 6.

8. Ibid., p. 8.

9. *Morbidity and Mortality Weekly Report.* Centers for Disease Control, Atlanta, August 2, 1991, vol. 40, no. 30, p. 514.

10. *Dallas Morning News,* "Free Vitamins May Ease Birth Defect Threat," September 18, 1992, p. 28A.

11. Angier, Natalie. "Vitamins Win Support as Potent Agents of Health." *New York Times,* March 10, 1992, p. C8.

12. Toufexis. "The New Scoop," pp. 56–57.

13. Nutritional Health Alliance (NHA) undated press release, p. 1.

14. David Kessler, personal communication.

15. Gerald Kessler, president NHA, personal communication.

16. Toufexis. "The New Scoop," p. 58.

17. Amendment to the Federal Food, Drug, and Cosmetics Act, Section 411, pp. 34–35.

18. Sugarman, Carole. "House Gives Vitamin Makers a Break," *Washington Post,* October 7, 1992, p. A15.

19. Mary Pendergast, personal communication.

20. *Washington Post*, September 14, 1993. p. A7.

21. *Grolier Encyclopedia.* "Orthomolecular Medicine."

22. *Grolier Encyclopedia.* "Megavitamin Therapy."

23. CRN voluntary dosage level recommendations.

24. Mary Pendergast, personal communication.

25. Transcript of speech by Michael R. Taylor, deputy commissioner for policy, FDA, at the annual conference of the Council for Responsible Nutrition, San Diego, September 21, 1992.

26. Brody, Jane E. "Vitamin E Greatly Reduces Risk of Heart Disease, Studies Suggest." *New York Times,* May 20, 1993, p. A1.

27. Ibid.

28. Ropp, Kevin L. "FDA to the Rescue." *FDA Consumer,* vol. 26, no. 28, October 1992, p. 14.

29. Ibid.

30. Ibid.

31. *Business Week,* "Why Gerber Is Standing Its Ground," March 17, 1986, p. 50.

32. Lewin, Tamar. "Tylenol Maker Finding New Crisis Less Severe." *New York Times,* February 12, 1986, p. B4. Kleinfeld, R. N. "Tylenol's Rapid Comeback." *New York Times,* September 17, 1983, p. B1.

33. *Report on the Contamination of Chilean Grapes with Cyanide in March 1989.* Undated. Chilean Exporters' Association, Moneda 920, Santiago, Chile, p. 33.

34. Ibid., p. 4.

35. Ibid., p. 7.

36. FDA Import Sample Summary, March 13, 1989, pp. 2–3.

37. Shenon, Philip. "Chilean Fruit Pulled from Shelves as U.S. Widens Inquiry on Poison." *New York Times,* March 14, 1989, p. A1.

38. Carlson, Margaret. "Do You Dare to Eat a Peach?" *Time,* March 27, 1989, p. 26.

39. Leary, Warren. "U.S. Urges Consumers Not to Eat Fruit from Chile." *New York Times,* March 14, 1989, p. A15.

40. Carlson. "Do You Dare," pp. 24–25.

41. Shenon, Philip. "U.S. Will Permit Fruit from Chile to Enter Market." *New York Times,* March 18, 1989.

42. Civil Action in the U.S. District Court for the Eastern District of Pennsylvania. Complaint, p. 50.

43. Ibid., pp. 50–51.

44. Ibid., p. 50.

45. FDA Import Sample Summary, p. 6.

46. Civil Action, pp. 53–54.
47. Ibid., pp. 54–55.
48. Ibid.
49. Ruszcyk, Brian. *The Ban on Chilean Fruit.* Independent survey, p. 20.
50. Civil Action, pp. 50–51.
51. *Table Grapes and Cyanide,* conclusions by the Crocker Nuclear Laboratory, University of California at Davis. February 10, 1992, p. 3.
52. Civil Action, pp. 54–55.
53. Civil Action, p. 58.
54. *Table Grapes and Cyanide,* p. 13.
55. FDA Internal Memorandum from HFR-MA150/David L. Chesney, DIB, PHI-DO.
56. FDA Import Sample Summary, p. 6.
57. Civil Action, p. 58.
58. Civil Action, p. 60. *Table Grapes and Cyanide,* pp. 12–13.
59. Sugarman, Carole. "Keeping the Faith." *Washington Post,* April 12, 1989. p. E1.

Chapter 7: The AIDS Gridlock

1. Hutt, Peter Barton. "Investigations and Reports Respecting FDA Regulation of New Drugs (Part II)." *Clinical Pharmacology and Therapeutics,* vol. 33, part II, 1983, p. 674.
2. Grabowski, Henry G., and John M. Vernon. *The Regulation of Pharmaceuticals.* American Enterprise Institute for Public Policy Research, Washington, D.C., 1983, p. 10.
3. Ibid.
4. Edgar, Harold, and David J. Rothman. "New Rules for Drugs: The Challenge of AIDS to the Regulatory Process." *Milbank Quarterly,* vol. 68, supplement I. Cambridge University Press, 1990, p. 113.
5. *Washington Post,* "FDA Recalls New Antibiotic." June 6, 1992, p. A5.
6. Kolata, Gina. "Drug Maker Didn't Heed Warning on Deadly Effect." *New York Times,* July 4, 1991, p. A1.
7. *Washington Post,* "Study Links Death Risk, Overuse of Asthma Drug," February 20, 1992, p. A3.
8. *Time,* "High Cost of Arthritis Relief," August 16, 1982, p. 47.
9. Arno, Peter S., and Karyn L. Feiden. *Against the Odds.* HarperCollins, New York, 1992, p. 4.
10. Butler, Samuel, *Erewhon.*

11. Sontag, Susan. *AIDS and Its Metaphors.* Farrar, Straus and Giroux, New York, 1988, p. 24.
12. Altman, Dennis. *AIDS In the Mind of America.* Anchor Press, Garden City, N. Y., 1986. p. 10.
13. Sontag, *Metaphors.* p. 26.
14. Arno and Feiden, *Against the Odds,* p. 90.
15. Ibid., p. 93.
16. Kolata, Gina. "In AIDS, Virus Forces Choice Between Longer Life or Eyesight." *New York Times,* December 8, 1987.
17. Arno and Feiden, *Against the Odds,* p. 160.
18. Ibid., p. 161.
19. Ibid., p. 212.
20. Wyss, Dennis. "The Underground Test of Compound Q." *Time,* October 9, 1989, p. 19.
21. Kolata, Gina. "Critics Fault Secret Effort to Test AIDS Drug." *New York Times,* September 19, 1989, p. C9.
22. Kwitny, Jonathan. *Acceptable Risks.* Poseidon Press, New York, 1992, p. 339.
23. *Economist,* "Queuing Up for the Unknown." June 30, 1990, p. 81.

Chapter 8: "Penicillin Couldn't Get Through That Fast"

1. Freudenheim, Milt. "Barr Seeks Generic AZT Approval." *New York Times,* April 19, 1991, p. D4. Wolfman, Brian. "Sleight-of-Hand in the AZT Patent Is Costly to the Consumer." *Public Citizen,* January/February, 1993, p. 25.
2. Arno, Peter S., and Karyn L. Feiden. *Against the Odds.* HarperCollins, New York, 1992, p. 39.
3. Ibid., p. 40.
4. Shilts, Randy. *And the Band Played On.* St. Martin's Press, New York, 1987, p. 594.
5. *Academic American Encyclopedia.*
6. Wachter, Robert M. *The Fragile Coalition: Scientists, Activists, and AIDS.* St. Martin's Press, New York, 1991, p. 84.
7. Kolata, Gina. "The Philosophy of the 'New FDA' is Mostly a Matter of Packaging." *New York Times,* May 19, 1991, p. E4.
8. FDA Press Office Update, October 19, 1992, p. 1.
9. Arno and Feiden, *Against the Odds,* pp. 271–272.
10. Kolata, Gina. "U.S. Is Asked to Control Prices of Drugs It Develops." *New York Times,* April 25, 1993, p. 36.
11. Ibid.

12. Ibid.
13. *New York Times*, "Judge Affirms Maker's Patent For Drug That Fights AIDS." July 23, 1993. p. A12.
14. Brown, David. "Study Questions AZT Ability to Slow AIDS." *Washington Post*, April 2, 1993, p. A1.
15. Ibid.
16. Ibid.
17. Ibid.
18. Ibid.

Chapter 9: Thicker Than Water

1. *Centers for Disease Control, HIV/AIDS Surveillance Report.* September 1–18, 1991, p. 8.
2. Rock, Andrea. "Inside the Billion-Dollar Business of Blood." *Money,* March 1986, p. 153.
3. Ibid., p. 158.
4. Feldschuh, Joseph, with Doran Weber. *Safe Blood.* Free Press, New York, 1990, pp. 55–56.
5. Shilts, Randy. *And the Band Played On.* St. Martin's Press, New York, 1987, p. 170.
6. Ibid., p. 171.
7. Feldschuh, *Safe.*
8. Ibid., p. 66.
9. Ibid.
10. Ibid.
11. Engleman, Edgar. "AIDS and the Blood Supply." *Report of the Presidential Commission on the Human Immunodeficiency Virus Epidemic,* No. 0–214–701. U.S. Government Printing Office, Washington, D.C., p. 4.
12. Thompson, Larry. "Red Cross Closing D.C. Blood Bank." *Washington Post,* February 27, 1990, p. B1.
13. Gaul, Gilbert M. "Blood Center in New York Will Lose License." *Philadelphia Inquirer,* May 11, 1990, p. A3.
14. Hilts, Philip J. "Blood Bank Faces Oregon Shutdown." *New York Times,* April 18, 1991, p. A23.
15. Hilts, Philip J. "Red Cross Faulted on Tainted Blood Reports." *New York Times,* July 11, 1990, p. B6.
16. Feldschuh, *Safe.* p. 112. Transcript of *60 Minutes,* Meredith Vieira, p. 13.
17. FDA Inspection Report, May 21, 1991.

18. House of Representatives, Committee on Energy and Commerce, background memorandum, p. 2.

19. Ibid.

20. Joseph Feldschuh, personal communication.

21. Jay Epstein, personal communication.

22. Ibid.

23. Ibid.

24. Gerald Sandler, personal communication.

25. Anonymous, personal communication.

26. Ibid.

27. Ibid.

28. Ibid.

29. Hearing before the Subcommittee on Oversight and Investigations of the Committee on Energy and Commerce, House of Representatives, Serial No. 101-169, July 13, 1990, U.S. Government Printing Office, Washington, D.C., p. 15.

30. Rosenthal, Andrew. "Protest on AIDS." *New York Times,* September 3, 1991, p. A20.

31. *New York Times,* letter to the editor, September 24, 1991, p. A30.

32. Hearing, p. 10.

33. Gerald Sandler, personal communication.

34. Statement by Gerald V. Quinnan, Jr., M.D., Acting Director, Center for Biologics Evaluation and Research, FDA, before the Committe on Energy and Commerce, Subcommittee on Oversight and Investigations, House of Representatives.

Chapter 10: Eat It, It's Good for You!

1. *Business Week,* "The Great American Health Pitch," October 9, 1989, p. 115.

2. Staff report prepared for the use of the Subcommittee on Oversight and Investigations of the Committee on Energy and Commerce, U.S. House of Representatives, September 1991, U.S. Government Printing Office, Washington, D.C., 102-N, pp. 2-3.

3. Ibid., p. 4.

4. Ibid., p. 5.

5. GAO-HRD-84-61, September 26, 1984.

6. Staff report, pp. 10-11.

7. GAO-RCED-83-153, September 9, 1983, p. 15.

8. Staff report, p. 19.

9. Hearings before a Subcommittee of the Committee on Government

Operations, House of Representatives, December 10, 1987, U.S. Government Printing Office, Washington, D.C., hereafter called Hearings 1987.

10. Hearings 1987, Appendix I.
11. Ibid., pp. 136–137.
12. Ibid., pp. 139–140.
13. Executive Order 12291.
14. Hearings 1987, p. 140.
15. Hearings before the Human Resources and Intergovernmental Relations Subcommittee of the Committee on Government Operations, House of Representatives, October 31 and November 9, 1989, U.S. Government Printing Office, Washington, D.C., hereafter called Hearings 1989, p. 6.
16. *Business Week,* p. 120.
17. Ibid., p. 114.
18. Ibid., p. 115.
19. Hearings 1989, pp. 7–8.
20. *Business Week,* p. 120.
21. Ibid., p. 119.
22. Ibid., pp. 119–120.
23. Ibid., p. 114.
24. Ibid., p. 115.
25. Hearings 1987, p. 142.
26. Ibid., p. 30.
27. Ibid., Appendix I.
28. Ibid., p. 34.
29. Ibid.
30. Ibid., p. 59.
31. Hearings 1989, p. 17.
32. Ibid.

Chapter 11: The Wunderkind

1. Sidney Wolfe, personal communication.
2. Gladwell, Malcolm. "Young Abruptly Resigns as Commissioner of Troubled FDA." *Washington Post,* November 14, 1989, p. A23.
3. Ibid.
4. Hilts, Philip J. "Under Revamping, FDA to Review Many Dormant Proposals on Safety." *New York Times,* June 10, 1991, p. A15.
5. Ibid.
6. Transcript of commencement address, Marshall-Wyeth Law School, May 16, 1993.

7. Ibid.

8. Ibid.

9. Gladwell, Malcolm. "FDA Chief Relishes Label of Lawman." *Washington Post,* October 24, 1991, p. A1.

10. *Barron's,* "Under a Microscope." March 2, 1992, p. 13.

11. Leary, Warren E. "Citing Lables, U.S. Seizes Orange Juice." *New York Times,* April 25, 1991, p. A18.

12. Leary, Warren E. "Company Agrees to Drop 'Fresh' from Name of Its Orange Juice." *New York Times,* April 27, 1991, p. A1.

13. Orvell, Miles. "The Metaphysics of Orange Juice." *New York Times,* July 1, 1991, p. A13.

14. *New York Times,* "Citing Labels," p. A18.

15. Burros, Marian. "FDA to Ban Product Labels That Mislead on Fat Content." *New York Times,* June 15, 1991, p. A1.

16. Ibid.

17. Sagon, Cindy. "Mister Kessler's Neighborhood." *Washington Post,* January 8, 1992, p. E1.

18. *New York Times,* "On the Spinach Front," January 20, 1992, p. A24.

19. Gladwell. "FDA Chief Relishes."

20. David Kessler, personal communication.

21. Ibid.

22. Gladwell. "FDA Chief Relishes."

23. *Wall Street Journal,* December 20, 1991, p. A1.

24. Ibid.

25. Hunt, Liz. "OMB Accused of Weakening Food-Labeling Proposals." *Washington Post,* September 26, 1991, p. A9.

26. Ibid.

27. Ibid.

28. Ibid.

29. *New York Times,* editorial, February 25, 1992.

30. Burros, Marian. "In Switch, FDA Offers Looser Rules on Labels." *New York Times,* February 11, 1992, p. A16.

31. David Kessler, personal communication.

32. *New York Times,* editorial, November 18, 1992.

33. Gladwell. "FDA Chief Relishes."

34. *New York Times,* August 11, 1992, p. A1.

35. *New York Times,* October 8, 1992, p. A1.

36. Anonymous, personal communication.

37. *New York Times,* October 8, 1992, p. A1.

38. *Washington Post,* October 8, 1992, p. A1.

Chapter 12: Commencement Day

1. Anonymous, personal communication.
2. Patrick, William. *The Food and Drug Administration.* New York: Chelsea House, 1988, p. 7.
3. "Promising Medical Devices to Be Speeded to Market." *Washington Post,* June 25, 1993, p. A2.
4. "Hill Panel Faults FDA Unit That Approves Medical Devices." *Washington Post,* June 2, 1993, p. A3.
5. Ibid.
6. *Washington Post,* editorial, January 23, 1993, p. A18.
7. Arno, Peter S., and Karyn L. Feiden. *Against the Odds.* New York: HarperCollins, 1992, p. 222.
8. David Kessler, personal communication.
9. Ibid.
10. Schwartz, John. "FDA Seeks More Data on Side Effects." *Washington Post,* June 2, 1993, p. A3.
11. Ibid.
12. Anonymous, personal communication.
13. Ibid.

REFERENCES

Altman, Dennis. *AIDS in the Mind of America.* Garden City, N.Y.: Anchor Press, 1986.

Arno, Peter S., and Karyn L. Feiden. *Against the Odds: The Story of AIDS Drug Development, Politics, and Profits.* New York: HarperCollins, 1992.

Blair, John M. *Economic Concentration.* New York: Harcourt Brace Jovanovich, 1972.

Blum, Richard, et al. *Pharmaceuticals and Health Policy.* London: Croom Helm, 1981.

Corea, Gina. *The Invisible Epidemic.* New York: HarperCollins, 1992.

Cray, William C., and C. Joseph Stetler. *Patients in Peril.* 1991.

Feldschuh, Joseph, with Doron Weber. *Safe Blood.* New York: Free Press, 1990.

Fontenay, Charles L. *Estes Kefauver: A Biography.* Knoxville: University of Tennessee Press, 1980.

Gorman, Joseph Bruse. *Kefauver: A Political Biography.* New York: Oxford University Press, 1971.

Grabowski, Henry G., and John M. Vernon. *The Regulation of Pharmaceuticals.* Washington, D.C.: American Enterprise Institute for Public Policy Research, 1983.

Greenfield, Jerome. *Wilhelm Reich vs. the U.S.A.* New York: Norton, 1974.

Grossman, Karl. *The Poison Conspiracy.* Sag Harbor, N.Y.: Permanent Press, 1983.

Hausman, Patricia. *The Right Dose.* Emmaus, Penn.: Rodale Press, 1987.

Joseph, Stephen C. *Dragon Within the Gates.* New York: Carroll and Graf, 1992.

Kwitny, Jonathan. *Acceptable Risks.* New York: Poseidon Press, 1992.

Lindsay, Cotton M. (ed.). *The Pharmaceutical Industry: Economics, Performance, and Government Regulation.* New York: Wiley, 1978.

Mintz, Morton. *The Therapeutic Nightmare.* Boston: Houghton Mifflin, 1965.

Mullan, Fitzhugh. *Plagues and Politics.* New York: Basic Books, 1989.

Nussbaum, Bruce. *Good Intentions: How Big Business and the Medical Establishment Are Corrupting the Fight Against AIDS.* New York: Atlantic Monthly Press, 1990.

Panem, Sandra. *The AIDS Bureaucracy.* Cambridge, Mass.: Harvard University Press, 1988.

Patrick, William. *The Food and Drug Administration.* New York: Chelsea House, 1988.

Pekkanen, John. *The American Connection.* Chicago: Follett, 1973.

Reekie, W. Duncan, and Michael H. Weber. *Profits, Politics, and Drugs.* London: Macmillan, 1979.

Reeves, Thomas C. *A Question of Character.* New York: Free Press, 1991.

Root-Bernstein, Robert. *Rethinking AIDS.* New York: Free Press, 1993.

Schlesinger, Arthur M. *A Thousand Days: John F. Kennedy in the White House.* Boston: Houghton Mifflin, 1965.

Schwartzman, David. *Innovation in the Pharmaceutical Industry.* Baltimore: Johns Hopkins University Press, 1976.

Shilts, Randy. *And The Band Played On.* New York: St. Martin's Press, 1987.

Silverman, Milton, and Philip R. Lee. *Pills, Profits, and Politics.* Berkeley: University of California Press, 1974.

Silverman, Milton, et. al. *Bad Medicine.* Stanford: Stanford University Press, 1992.

Silverman, Morton. *The Drugging of the Americas.* Berkeley: University of California Press, 1976.

Smith, Mickey C. *Principles of Pharmaceutical Marketing.* Philadelphia: Lea and Febiger, 1983.

Sobol, Richard B. *Bending the Law.* Chicago: University of Chicago Press, 1991.

Sontag, Susan. *Illness as Metaphor.* New York: Farrar, Straus and Giroux, 1977.

——. *AIDS and Its Metaphors.* New York: Farrar, Straus and Giroux, 1988.

Teitelman, Robert. *Gene Dreams.* New York: Basic Books, 1989.

Tuchman, Barbara. *The March of Folly.* New York: Knopf, 1984.

Wachter, Robert M. *The Fragile Coalition: Scientists, Activists, and AIDS.* New York: St. Martin's Press, 1991.

White, Ryan. *My Own Story.* New York: Dial, 1991.

Wolfe, Sidney M. et al. *Worst Pills, Best Pills.* Washington, D.C.: Public Citizen Health Research Group, 1988.

Wood, Glenn G., and John E. Deitrich. *The AIDS Epidemic: Balancing Compassion and Justice.* Portland, Oreg.: Multnomah Press, 1990.

Young, James Harvey. *Pure Food.* Princeton, N.J.: Princeton University Press, 1989.

——. *The Medical Messiahs.* Princeton, N.J.: Princeton University Press, 1967.

——. *The Toadstool Millionaires.* Princeton, N.J.: Princeton University Press, 1961.

INDEX